The Appearance of Human Skin: A Survey

The Appearance of Human Skin: A Survey

Takanori Igarashi

Kao Corporation
Tokyo, Japan
igarashi.takanori@kao.co.jp

Ko Nishino

Drexel University
Philadelphia, PA
kon@cs.drexel.edu

Shree K. Nayar

Columbia University
New York, NY
nayar@cs.columbia.edu

the essence of knowledge

Boston – Delft

Foundations and Trends® in Computer Graphics and Vision

Published, sold and distributed by:
now Publishers Inc.
PO Box 1024
Hanover, MA 02339
USA
Tel. +1-781-985-4510
www.nowpublishers.com
sales@nowpublishers.com

Outside North America:
now Publishers Inc.
PO Box 179
2600 AD Delft
The Netherlands
Tel. +31-6-51115274

The preferred citation for this publication is T. Igarashi, K. Nishino and S. K. Nayar, The Appearance of Human Skin: A Survey, Foundations and Trends® in Computer Graphics and Vision, vol 3, no 1, pp 1–95, 2007

ISBN: 978-1-60198-086-1

Editorial Scope

Foundations and Trends® in Computer Graphics and Vision will publish survey and tutorial articles in the following topics:

- Rendering: Lighting models; Forward rendering; Inverse rendering; Image-based rendering; Non-photorealistic rendering; Graphics hardware; Visibility computation
- Shape: Surface reconstruction; Range imaging; Geometric modelling; Parameterization;
- Mesh simplification
- Animation: Motion capture and processing; Physics-based modelling; Character animation
- Sensors and sensing
- Image restoration and enhancement
- Segmentation and grouping
- Feature detection and selection
- Color processing
- Texture analysis and synthesis
- Illumination and reflectance modeling
- Shape Representation
- Tracking
- Calibration
- Structure from motion
- Motion estimation and registration
- Stereo matching and reconstruction
- 3D reconstruction and image-based modeling
- Learning and statistical methods
- Appearance-based matching
- Object and scene recognition
- Face detection and recognition
- Activity and gesture recognition
- Image and Video Retrieval
- Video analysis and event recognition
- Medical Image Analysis
- Robot Localization and Navigation

Information for Librarians

Foundations and Trends® in Computer Graphics and Vision, 2007, Volume 3, 4 issues. ISSN paper version 1572-2740. ISSN online version 1572-2759. Also available as a combined paper and online subscription.

Foundations and Trends® in
Computer Graphics and Vision
Vol. 3, No. 1 (2007) 1–95

DOI: 10.1561/0600000013

The Appearance of Human Skin: A Survey

Takanori Igarashi[1], Ko Nishino[2] and Shree K. Nayar[3]

[1] *Skin Care Research Laboratory, Kao Corporation, Tokyo, Japan, igarashi.takanori@kao.co.jp*
[2] *Department of Computer Science, Drexel University, Philadelphia, PA, kon@cs.drexel.edu*
[3] *Department of Computer Science, Columbia University, New York, NY, nayar@cs.columbia.edu*

Abstract

Skin is the outer-most tissue of the human body. As a result, people are very aware of, and very sensitive to, the appearance of their skin. Consequently, skin appearance has been a subject of great interest in various fields of science and technology. In particular, research on skin appearance has been intensely pursued in the fields of computer graphics, computer vision, cosmetology, and medicine. In this survey, we review the most prominent results related to skin in these fields and show how these seemingly disconnected studies are related to one another. In each of the fields, the optical behaviors of specific skin components have been studied from the viewpoint of the specific objectives of the field. However, the different components of skin produce different types of optical phenomena that are determined by their physio-anatomical characteristics (sizes, shapes, and functions of the components). The final appearance of skin has contributions from complex optical interactions of many different skin components with light. In order to view these interactions in a unified manner, we describe and categorize past works based on the physiological and anatomical characteristics of the various skin components.

Contents

1

Why is Skin Appearance Important?

Skin is the outermost tissue of the human body. As a result, people are very aware of, and very sensitive to, the appearance of their skin. Consequently, skin appearance has been a subject of great interest in several fields of science and technology. As shown in Figure 1.1, research on skin appearance has been intensely pursued in the fields of medicine, cosmetology, computer graphics, and computer vision. Since the goals of these fields are very different, each field has tended to focus on specific aspects of the appearance of skin. The goal of this study is to present a comprehensive survey that includes the most prominent results related to skin in these different fields and show how these seemingly disconnected studies are closely related.

In the field of computer graphics, computational modeling of the appearance of skin is today considered to be a very important topic. Such skin appearance models are widely used to render fictional human characters in movies, commercials, and video games. For these "virtual actors" to appear realistic and be seamlessly integrated into a scene, it is crucial that their skin appearance accurately captures all the subtleties of actual human skin under various viewing and lighting conditions. Although great progress has been made in making rendered skin

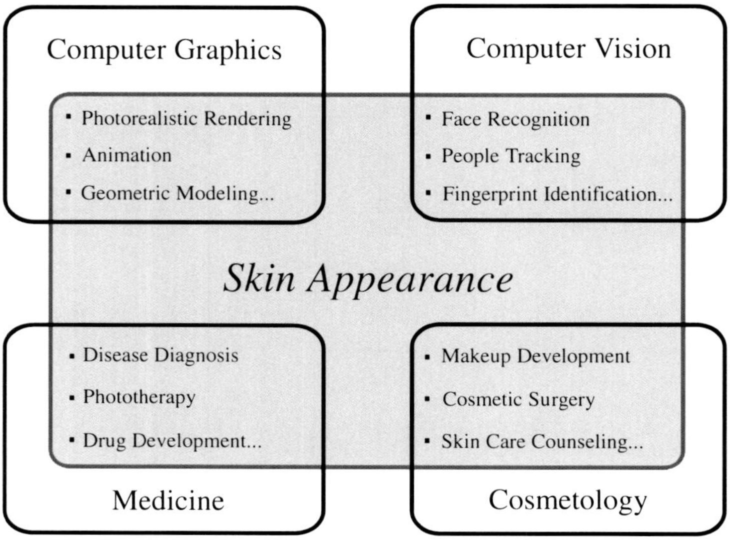

Fig. 1.1 Many different research fields have conducted extensive research on the appearance of skin. The fields of medicine, cosmetology, computer graphics, and computer vision have been most active in the study of skin appearance. Studies in each of these fields provide us knowledge and insights regarding different aspects of skin appearance.

appear realistic, it is still far from perfect and is easily recognized as being rendered rather than real. In short, a computationally efficient and yet realistic skin model remains an open problem in computer graphics.

In computer vision, a detailed and accurate model of skin appearance is of great value in identifying individuals. For instance, human identification based on fingerprints has made substantial progress and is now a widely used biometric technology. It is now widely acknowledged that accurate models of the appearance of skin in other parts of the body could be useful for human identification as well. For instance, technologies that recognize the pattern of blood vessels in the palm and the finger have been recently developed and have shown good performance in identification. In order to reliably exploit similar signatures of skin appearance from other body regions, we need a more comprehensive understanding of the optical characteristics of skin.

Skin also has aesthetic relevance. The desire to have beautiful and healthy looking skin has been a centuries-old quest for humans. Skin with brighter complexion and smoother surface tends to be perceived as being healthier and more attractive. Making skin appear beautiful is the primary goal of cosmetology. For instance, foundations are widely used to hide skin imperfections and make skin look younger. Despite the enormous investments made in skin research, today's foundations are far from perfect. While they may hide imperfections and make skin appear more uniform, the final appearance of skin coated with a foundation always has an artificial look to it. Recently, skin counseling systems have been developed to help a person identify cosmetic products that would be most suited to them. Such systems can also benefit from more accurate and detailed models of skin appearance.

Needless to say, the appearance of skin is of vital importance to the field of medicine. During the diagnosis of skin diseases, careful observation and assessment of the appearance of the diseased area is always the first and most important step. Recently, photo-diagnosis and photo-therapy have become popular methods for treating skin diseases. In these techniques, light is used to detect and treat lesions in the skin. Such techniques are non-invasive and hence patients are not subjected to pain and scars during the treatment. In order to increase the precision of such systems, we need more precise models of the interaction of light with dermal tissues.

In this survey, we will summarize and relate studies on skin appearance conducted in the above fields. Our goal is to present the disconnected works in these different areas within a single unified framework. In each of the above fields, the optical behaviors of specific skin components have been studied from the viewpoint of the specific objectives of the field. However, the different components of skin produce different types of optical phenomena that are determined by their physio-anatomical characteristics (sizes, shapes, and functions of the components). The final appearance of skin has contributions from complex optical interactions of many different skin components with light. In order to view these interactions in a unified manner, it is meaningful to describe and categorize past works based on the physiological and anatomical characteristics of the various skin components. To this end, we will first outline the physio-anatomical characteristics of skin that are important to its appearance. Then, we will review previous studies that have been conducted on each of the structural components of skin.

We will start our survey by describing the basic functions of human skin in Section 1.1. This knowledge is necessary to understand the physio-anatomical properties of the components of skin. In Section 1.2, we will propose a taxonomy of skin appearance that serves as the basic structure of our survey. In this taxonomy, we summarize the important physio-anatomical components of skin and the optical phenomena they produce. In Chapter 2, we will describe in detail the physio-anatomical structure and character of each skin component. In Chapter 3, we will review studies on skin appearance that have been conducted in the four fields shown in Figure 1.1. We hope our survey will have two effects. The first is to broaden and deepen the reader's understanding of skin appearance. The second is to spur new interdisciplinary research on skin appearance.

1.1 What is Skin?

Skin is the outermost tissue of the body and the largest organ in terms of both weight and surface area. It has an area of approximately $16{,}000\,\mathrm{cm}^2$ for an adult and represents about 8% of the body weight. As seen in Figure 1.2, skin has a very complex structure that consists of many components. Cells, fibers and other components make up several different layers that give skin a multi-layered structure. Veins, capillaries and nerves form vast networks inside this structure. In addition, hairs stick out from the inside of skin. Numerous fine hair furrows are scattered over the surface of skin.

Skin performs a wide variety of functions resulting from chemical and physical reactions inside these components. The major function of skin is to act as a barrier to the exterior environment. It protects the body from friction and impact wounds with its flexibility and toughness. Harmful chemicals, bacteria, viruses, and ultraviolet light are also prevented from entering the body by the skin. It also prevents water loss and regulates body temperature by blood flow and evaporation of sweat. These functionalities are critical to our well-being. The secretion

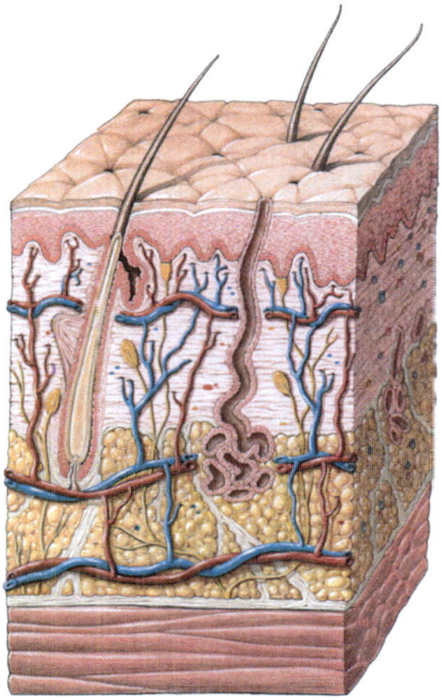

Fig. 1.2 A cross-sectional schematic diagram of skin. Skin is a complex multi-layered tissue consisting of various types of components, including veins, capillaries, hairs, cells, fibers, etc. (Image courtesy of A.D.A.M.)

of sweat and skin lipid cause the elimination of a number of harmful substances resulting from metabolic activities in the intestines and the liver. Furthermore, skin has a large number of nerve fibers and nerve endings that enable it to act as a sensory organ. When skin is exposed to sunlight, it can produce vitamin D, a vital chemical substance for the body [144].

These functions of skin tend to vary in degrees according to age, race, gender, and individual. For instance, older skin tends to lose its flexibility and toughness because the structure of skin slowly denatures with age. The light-protection ability of skin among different races varies due to the differences in the volume of melanin which absorbs ultraviolet light. These functional differences are in most cases a result of physio-anatomical variations within the structure of skin. It is these physio-anatomical variations that lead to the diverse appearances of skin. Hence, in order to understand the appearance of skin, it is crucial to understand the physiology and anatomy of skin. In the next section, we will present a taxonomy of skin appearance that is based on physiology and anatomy. Then, in Chapter 2, we will use this taxonomy to describe the physio-anatomical properties of the various skin components.

1.2 Taxonomy of Human Skin Appearance

In order to understand the appearance of skin, it is important to understand the optical/visual properties of the constituent components of skin. In this section, we present a hierarchical representation of skin components that is based on the scale of the optical processes induced by the components.

As shown in Figure 1.3, the components of skin appearance can be categorized along an axis that represents spatial scale. Here, we only focus on skin components that have a measurable contribution to its appearance. We refer to the smallest components as *microscale*, larger components as *mesoscale*, and the largest components as *macroscale*. Each scale is subdivided into finer levels based on physiology and anatomy. As a result, skin can be viewed as a hierarchical organ in which components at one level serve as building blocks to constitute higher-level components. The components in each level have their own visual properties. Each of these visual properties is studied based on its underlying physical phenomena. The scattering or appearance model that describes these phenomena are listed in the rightmost column in Figure 1.3.

1.2.1 Microscale

Cellular-level elements and *skin layers* constitute the finest scale of the physio-anatomical structure of skin. The sizes of these components are very small and they are barely visible to the naked eye. The visual properties of these elements are the result of their optical interactions with incident light. From an optical viewpoint, the dominant effects produced at this scale are scattering and absorption. These effects vary depending on the sizes, shapes, and optical parameters such as the refractive indices of the elements. For example, fibers and organelles

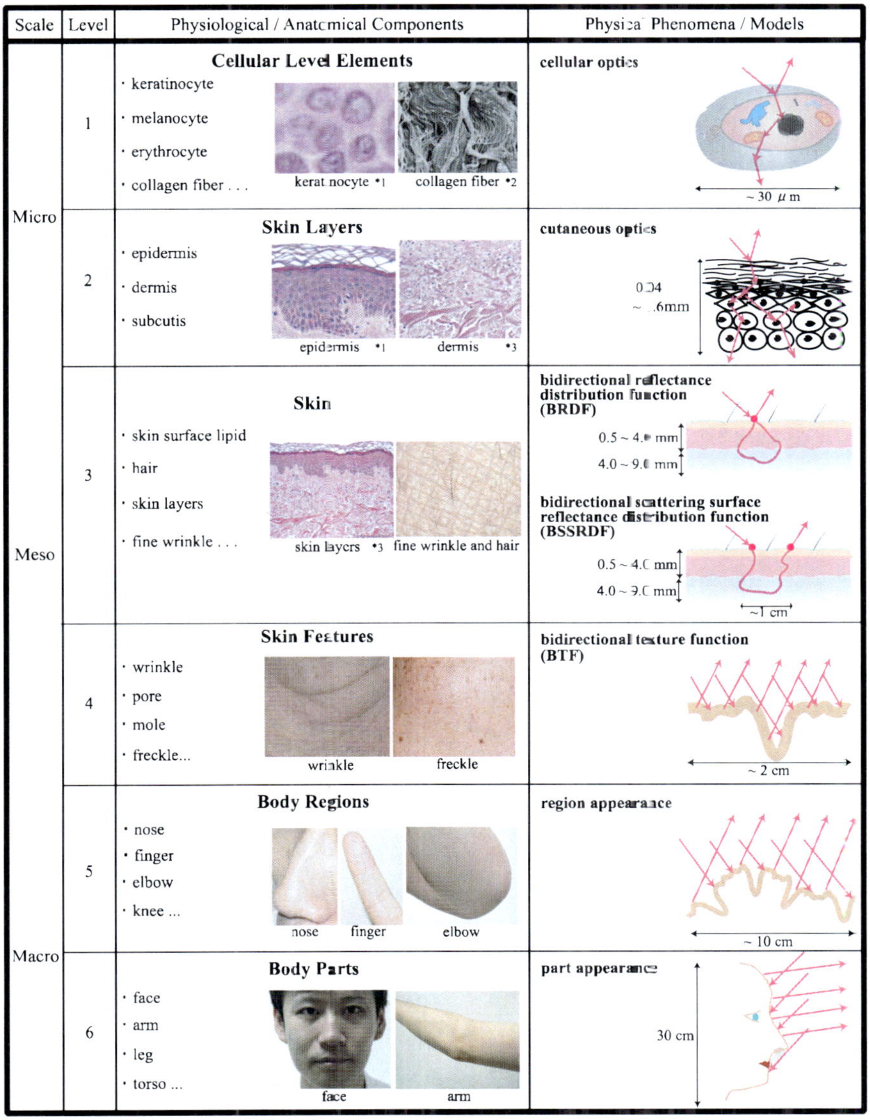

Fig. 1.3 Taxonomy of the appearance of skin. The components of skin appearance can be hierarchically categorized along an axis that represents physical scale. We review the studies on skin appearance done in different fields based on this taxonomy. *1 Photo courtesy of Chrustopher Shea, MD, Duke University Medical Center; *2 Photo from Nanoworld Image Gallery, Centre for Microscopy and Micronanoanalysis, The University of Queensland; *3 Photo courtesy of T. L. Ray, MD, University of Iowa College of Medicine.

that are found in cells behave as strong scatterers. The cell membranes and the blood vessel walls behave like reflectors and refractors. The aggregation of these optical phenomena determine the visual properties of the components at higher levels. Some of these elements, such as organelles, have sizes close to the wavelength of visible light. Hence, the optical properties at this level must be studied using wave optics.

The cellular-level elements constitute several primary layers of skin: epidermis, dermis, and subcutis, which are classified in Level 2 in Figure 1.3. These layers have very different structures and constituents and hence their physiological characteristics are different from each other. For example, the epidermis is a very thin layer (0.2 mm on average) which mainly consists of cells. On the other hand, the dermis is a thick layer (2 mm on average) composed of more fibers compared to the epidermis. These physio-anatomical differences have large influences on the light propagation in these layers and lead to very different optical effects. For example, the epidermis is a transparent optical medium and the dermis is a turbid medium. These optical differences enable us to view these layers as the primary optical media for describing the optical properties of higher-level components.

1.2.2 Mesoscale

Skin and *skin features* constitute the mesoscale. At this scale, the components become visible to the naked eye. The visual properties of these components are mainly determined by the optical phenomena that are induced by finer-scale components — components in the microscale.

Skin, as categorized in Level 3 in Figure 1.3, is composed of *skin layers*, *skin surface lipid*, *hairs*, *fine wrinkles*, etc. The appearance of skin can be viewed as the combined effect of the optical phenomena induced by these substructures. Skin layers include the lower-level components in Level 2 — epidermis, dermis, and subcutis. Visual property of skin layers can also be considered as the combined effect of the optical events that take place in each of these layers. Hence, understanding the optical properties in the microscale is crucial to understand the visual properties of the components in the mesoscale.

Skin surface lipids, hairs, and fine wrinkles are found on the surface of skin. They contribute interesting optical effects. For example, the appearance of skin after sweating usually becomes more glossy. This change of appearance is mainly due to the reflection of incident light by the film of skin surface lipids. The appearance of skin with dense hair and fine wrinkles tends to be more matte. This is because of the additional scattering of incident light by the hairs and fine wrinkles.

Skin constitutes higher level components – skin features such as freckles, moles, wrinkles, and pores (see Level 4 in Figure 1.3). These features can be viewed as morphological variations of skin. For example, freckles and moles tend to produce two-dimensional variation in skin color. In contrast, wrinkles cause deep furrows and flat planes and are inherently three-dimensional textures. Hence, the visual properties of skin features are influenced by not only the optical properties of the skin layers but also the morphology of skin.

1.2.3 Macroscale

Body regions and *body parts* are classified as macroscale and physiologically assigned to Level 5 and Level 6, respectively (see Figure 1.3). The appearance of skin varies across different regions of the body. This is because the physio-anatomical characteristics of the lower-level components can differ significantly from one region of the body to another. For example, the nose and the forehead have greater amounts of skin surface lipid compared to the cheek. As a result, the nose and the forehead tend to appear more glossy than the cheek. To our knowledge, there are no physical models that describe these appearance variations over the body in a unified framework. Body parts such as the face, arm, leg, and torso are clusters of body regions. The appearance of each body part includes the appearances of the body regions that constitute it. Again, we are not aware of any physical or empirical model for describing part appearances in a unified manner.

It is interesting to note that the four fields that have been involved in skin research have tended to focus on different scales or levels of skin appearance. In computer graphics and computer vision, components in the visible scale have been studied. This is because the main objectives in these fields are to render and recognize skin appearance. Thus, previous work in graphics and vision provide us with knowledge about the visual properties of skin mainly at the meso and macroscales. On the other hand, research in medicine has focused on lower-level elements. This is because skin diseases are usually caused by disorders in the microscale components. Thus, past work in medicine provides us with knowledge about the optical properties of skin at the microscale. By reviewing work in these different fields, we can span all the scales of skin appearance and, at the same time, describe all of the previous works in a consistent manner.

2

Physiology and Anatomy of Human Skin

The optical and visual properties of skin components at each level of our taxonomy (Figure 1.3) differ significantly depending on their physio-anatomical characteristics. For example, light scattering behaviors of cells and fibers (Level 1) depend on their sizes and shapes. Light propagation in the skin layers, i.e., the epidermis and the dermis (Level 2), are very different since their structures, densities, and thicknesses vary greatly. Reflection at the surface of skin is influenced by the morphological characteristics of fine wrinkles (Level 3). The appearance of wrinkles themselves also depend on their morphological characteristics such as depth, width, and density variations. Most of these optical and visual properties are different for different body regions (Level 5) and body parts (Level 6) since the physio-anatomical characteristics of the lower-level components (Levels 1 to 4) vary across the body. The above examples are used to convey the importance of understanding the physio-anatomical properties of each of the skin components.[1]

[1] Readers are referred to [144] for an entry point to the rich general literature on the physiology of human skin.

2.1 Level 1: Cellular-Level Elements

Although skin is composed of various types of cellular-level elements, *cells*, *fibers*, and *chromophores* are of special relevance to us. This is because light scattering and absorption in these fundamental elements are the building blocks of the gross optical phenomena observed at the cellular level.

2.1.1 Cells

Skin includes various types of cells. The main cells are *keratinocyte*, *fibroblast*, *fat cell*, *melanocyte*, and *erythrocyte*. These cells are present in different locations and have different structures and functions.

Keratinocytes are the most abundant cells in the epidermis. These cells produce fibriform proteins called keratin which contribute to the rigidity of the outermost layer of skin. Keratinocytes protect the body from the external environment, for instance from stimulation, friction, and viruses, while retaining moisture. Keratinocytes can be further categorized into four types of cells based on their functions, and structures: *basal cells*, *prickle cells*, *granular cells*, and *horny cells*. Although these cells have the same origin, they have different shapes, functions, and subcellular-level elements called organelles. For example, the basal cell, which reproduces keratinocytes, is a cylindrical and soft living cell. On the other hand, the horny cell, which mainly acts as a protector from the external environment, is a very flat and hard dead cell in which most organelles are degenerate.

Fibroblasts are long and narrow cells present in the dermis, the second skin layer beneath the epidermis. They produce collagen and elastin fibers which are the primary constituents of the dermis.

Fat cells are the most abundant cells of the dermis. These cells accumulate fat and their sizes vary according to the volume of fat contained in them. These cells do not absorb much light. On the other hand, *melanocyte* and *erythrocyte* cells, both of which contain chromophores, mainly absorb light.

Melanocytes carry *melanin* which is one of the main light-absorbing pigments in skin. There are generally 1000–2000 melanocytes in 1 mm^2 of skin. This cell contains specialized organelles called melanosomes. When skin is exposed to sunlight, melanosomes are activated and melanin is produced. The density of melanosomes depends on the body region. For example, regions that are frequently exposed to sunlight, such as the face, have higher density than other regions.

Erythrocytes (or **red blood cells**) are the carriers of hemoglobin, another light-absorbing pigment in skin. As seen from Figure 2.1, erythrocytes have biconcave structures. The diameter of an erythrocyte is approximately 5 μm [3]. Erythrocytes usually contain more than 300 mg/mL of hemoglobin and carry oxygen from the lungs to tissues and carbon dioxide from tissues to the lungs.

As shown in Figure 2.2, a typical cell is composed of a cell membrane and organelles such as nucleus, mitochondria, lysosome, cytoplasm, Goldi apparatus, endoplasmic reticulum, etc. The nucleus, which is the largest spherical organelle, ranges from 3 to 10 μm in size and is enclosed in a membrane called the nuclear envelope. The nucleus includes most of the DNA of a cell and serves as the storage area for genetic information. The mitochondria is 0.5–1.5 μm in size and is an oval-shaped organelle composed of a double membrane. The mitochondria generates

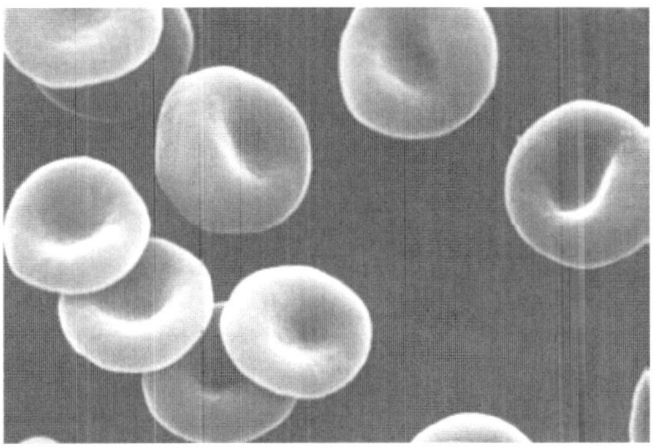

Fig. 2.1 Scanning electron microscope image of erythrocytes. Erythrocytes contain one of the main light-absorbing chemical compounds in skin, *hemoglobin*. (Courtesy of A. M. Cohen, Tel-Aviv University.)

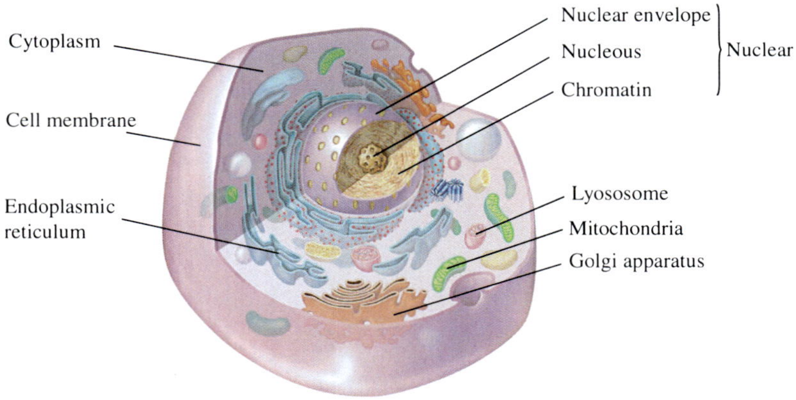

Fig. 2.2 A schematic representation of a cell. The cell usually contains many types of organelles. In some types of cells, such as the horny cell, these organelles are denatured. (Image courtesy of A.D.A.M.)

energy from food. The cell membrane is the outermost layer of a cell and has a doubly layered structure of lipids (bilayer membrane). The thickness of the cell membrane is approximately 15 nm [49].

2.1.2 Fibers

Skin contains several types of fibers. We will consider *keratin*, *collagen*, and *elastin* as typical types of fibers found in skin.

Keratin fibers are mainly found in the outer-level epidermal cells, including horny cells. These fibers protect the inner side of skin from the external environment. At the same time, they contribute to moisture-retention in skin by holding water. The length and diameter of these fibers depend on the amount of moisture they actually hold.

Collagen fibers are the main constituents of the dermis. They represent about 70% of the dermis in dry weight. These fibers form vast and tough networks providing the dermis with strength, tension, and elasticity. As shown in Figure 2.3, a collagen fiber is 0.5–3 μm in diameter [3] and has a very long shape. A collagen fiber has a hierarchical structure [3] and is essentially a bundle of smaller microcables called *collagen fibrils*. Collagen fibrils are 10–300 nm in diameter and many micrometers long. Acollagen fibril is a bundle of triple stranded collagen

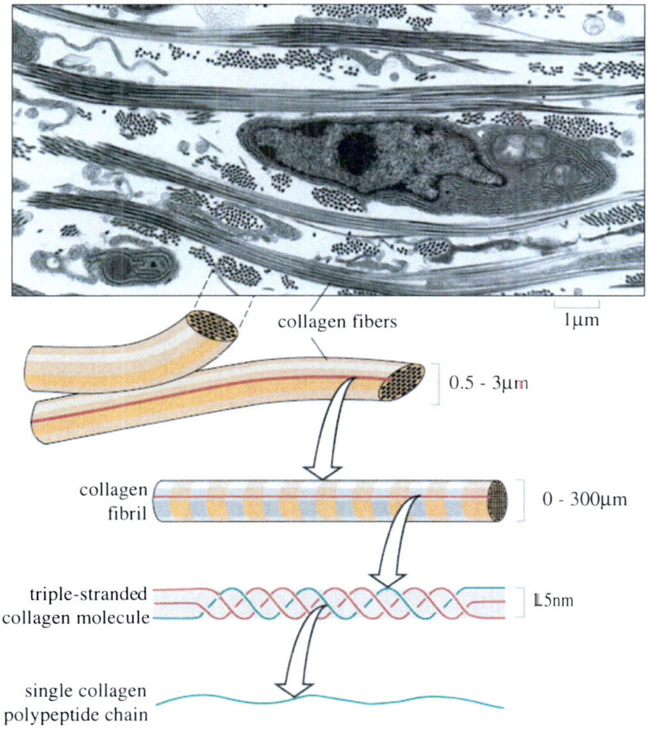

Fig. 2.3 The structure of collagen fibers. Collagen fibers are composed of collagen fibrils that are bundles of collagen molecules made of polypeptide chains. (Reprinted with permission [3]; ©1998 Garland Science Publishing.)

molecules (about 1.5 nm in diameter [3, 111] and 300 nm long [111]), three polypeptide chains that are wrapped around each other as a triple helix.

The structure of collagen fibers starts to deteriorate around the age of 30. Photo-damaging, which occurs with ultraviolet light in sunlight, and smoking also denature the structure of collagen fibers. These factors eventually cause morphological changes to the network of collagen fibers. This leads to loss of skin elasticity and finally induces wrinkling [2, 18, 28, 165, 101].

Elastin fibers are random coiled proteins that are also present in the dermis. These fibers are thinner than collagen bundles (1–3 μm in diameter [111]). They occupy 2%–4% of the total weight of the dermis. An elastin fiber consists of two components — micro-fibrils and matrix elastin. The micro-fibrils are aggregated at the periphery of elastic fiber (10–12 nm thick) and are also present within elastin fibers as strands aligned along the longitudinal direction (15–80 nm thick) [111].

Elastin fibers provide skin with elasticity and resilience. Even though the volume of elastin fibers is much smaller than that of collagen fibers, elastin fibers also play an important role in providing structural support to the dermis. Similar to collagen fibers, aging and ultraviolet light degrade elastin fibers, which finally leads to wrinkling. Elastin fibers are extensible and return to their original shapes after stretching. This property is not found in collagen fibers [111].

2.1.3 Chromophores

Skin includes various types of light-absorbing chemical compounds called chromophores. Among these chromophores, *melanin* and *hemoglobin* are especially important for understanding the appearance of normal skin, since they absorb light particularly in the visible wavelength range [7, 193].

Melanin is the dominant chromophore of the epidermis. It can also be found in hair. Melanin is first produced in melanosomes, then is diffused into the epidermal layer, and moves up toward the surface of skin while denaturing. Through this upward process, melanin changes its color from tan to white.

Melanin is divided into two types, *eumelanin* and *pheomelanin*, depending on its chemical structure. Eumelanin is a black or dark brown chromophore usually found in dark hair and eyes. Pheomelanin is yellow or reddish brown chromophore that is observed in red hair and feathers. Usually, normal skin contains both types of melanin. The physiological function of melanin is to protect the inside of skin by absorbing and scattering ultraviolet light. When exposed to sunlight, melanocytes start to produce melanin. This is the biological reaction that eventually makes our skin appear tanned.

The color of skin depends on the fraction of the volume of the melanosomes in the epidermis. In the light colored skin of Caucasians, the fraction is only between 1% and 3%. In the skins of well-tanned Caucasians and Mediterraneans, the percentage increases from 11% to 16%. In dark colored African skin, it goes up to 43% [82].

Hemoglobin is a red colored chromophore found in erythrocytes. Hemoglobin represents 95% of the dry mass of an erythrocyte. Hemoglobin binds oxygen effectively and carries oxygen to every body site through vessels and capillaries. When hemoglobin contains oxygen, it is called oxy-hemoglobin. Otherwise, it is called deoxy-hemoglobin. Usually, in the vein, more than 47% of the hemoglobin is oxy-hemoglobin [8]. Oxy-hemoglobin is a brighter shade of red than deoxy-hemoglobin.

2.2 Level 2: Skin Layers

Cellular-level elements (Level 1) form the three different skin layers: *epidermis*, *dermis*, and *subcutis*. These layers are composed of different types of cellular-level elements. Hence, they are very different in terms of structure and function. As a result, they exhibit different types of light propagation.

2.2.1 Epidermis

The epidermis is the outermost layer of skin. There are no veins and capillaries in this layer. Its thickness is about 0.2 mm on average and this thickness varies depending on the location on the body. Furthermore, the thickness also varies according to the volume of water that the epidermis holds.

The epidermis is further divided into five sublayers. From the bottom (innermost), these sublayers are *stratum basale* (*basal cell layer*), *stratum spinosum* (*prickle cell layer*), *stratum granulosum* (*granular cell layer*), *stratum lucidum* (*clear layer*), and *stratum corneum* (*horny cell layer*) (see Figure 2.4).

The epidermis is a metabolically active tissue. Keratinocytes produced in stratum basale move upward to the outer surface. This process is called *turn-over*. During this turn-over, keratinocytes change their structures and physiological functions. One cycle of this turn-over

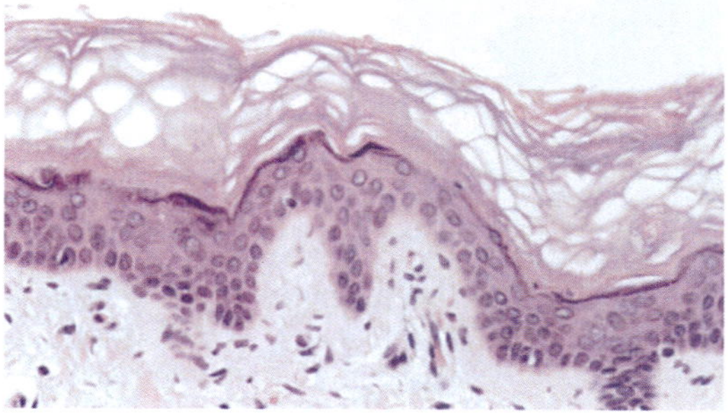

Fig. 2.4 A microscope image of the epidermis. The epidermis is composed of four sublayers: stratum corneum, stratum granulosum, stratum spinosum, and stratum basale. In the sole and the palm, there is an additional layer called stratum lucidum underneath stratum corneum. (Photo courtesy of Christopher Shea, Duke University Medical Center.)

process takes about 28 days. In the following, we will describe the physio-anatomical properties of each sublayer.

Stratum basale (basal cell layer) is the deepest sublayer of the epidermis and is composed of a single layer of basal cells. This sublayer forms the boundary to the dermis. Keratinocytes are produced in this sublayer. It holds approximately 8% of the water in the epidermis. With aging, this layer becomes thinner and loses the ability to retain water. Melanocytes, which were mentioned in the previous section, also lie in this layer.

Stratum spinosum (prickle cell layer) refers to the 10–20 layers that lie on top of the basal cell layer. Basal cells, through the process of turn-over, make their shape somewhat flatter (multi-sided) and form these layers. These cells are called prickle cells and have little spines on the outside of their membrane. The thickness of this sublayer is typically from 50–150 μm [7].

Stratum granulosum (granular cell layer) is composed of 2–4 granular cell layers. The typical thickness is 3 μm [7]. In this sublayer, cornification called keratinization of keratinocytes begins. In this process, organelles such as nuclei and mitochondria start to dissolve. Cells are increasingly filled with keratin fibers and contain less moisture than basal and prickle cell layers. The shape of these cells becomes much flatter during this process.

Stratum lucidum (clear layer) can only be found in soles and palms. It is a highly refractive sublayer. Its cells become flatter and more densely packed during turn-over.

Stratum corneum (horny cell layer) is the exterior sublayer of the epidermis. Its thickness ranges from 8–15 μm [7]. This sublayer is composed of several layers of hexagonal-shaped flat and hard cells named horny cells or corneocytes. These are dry dead cells without organelles and filled with keratin fibers. This sublayer prevents excessive dehydration of the skin tissue and usually contains 10%–15% of the mass of water in the epidermis, depending on the skin condition. Horny cells are surrounded by intercellular lipids. A principal constituent is *ceramide*, which plays a crucial role in water retention [108]. Horny cells also contain special chemical compounds called natural moisturizing factor (NMF) that also plays an important role in retaining skin moisture. NMF is composed of sodium PCA, sphinolipids and ceramides, phospholipids, fatty acids, glycerol, squalane, and cholesterol [108]. Skin that lacks NMF and ceramide tends to be very dry.

2.2.2 Dermis

The dermis is the second layer of skin, beneath the epidermal layer. This layer is much thicker than the epidermis (usually 1–4 mm [7]). The main components of the dermis are collagen and elastin fibers. Compared to the epidermis, there are much fewer cells and much more fibers in the dermis. Dermis has the following two sublayers.

Papillary layer is the upper sublayer of the dermis that is clearly demarcated from the epidermis. This sublayer is a loosely connected tissue and includes a large amount of nerve fibers, capillaries, water, and cells (e.g., fibroblasts). In this sublayer, collagen fibers form a finer network than those of the reticular layer [33].

Reticular layer constitutes the lower part of the dermis and represents a continuous transition to the subcutis. This sublayer has a denser and thicker network than the papillary layer and includes fewer nerve fibers and capillaries. In this sublayer, collagen fibers are aggregated into thick bundles which are mostly aligned parallel to the surface of skin [33].

The micro-anatomical complexity of skin, particularly in Level 1, makes quantitative analysis of the optical properties of skin difficult. However, it can be significantly simplified by considering the physio-anatomical characteristics of each skin layer. As mentioned above, the epidermis and the dermis are very different in composition, thickness, and functions. Hence, these two layers can be considered to be independent of each other in terms of optical behaviors. Indeed, the epidermis and the dermis are viewed as independent optical media in many early studies on skin optics. For example, melanin is present only in the epidermis. On the other hand, hemoglobin is found only in the dermis since there are no veins and capillaries in the epidermis. Hence, the epidermis can be essentially viewed as a *melanin layer* and the dermis can be viewed as a *hemoglobin layer* when analyzing the absorption properties of skin.

2.2.3 Subcutis

Subcutis, or hypodermis, is the third layer beneath the dermis. It is important to note that it is not categorized as another skin layer. Subcutis is an elastic layer and includes a large amount of fat cells that work as a shock absorber for blood vessels and nerve endings. The thickness of this layer is reported to be 4–9 mm on average. However, the actual thickness differs from person to person and also depends on the body region.

2.3 Level 3: Skin

Skin contains *skin layers*, *fine wrinkles*, *hairs*, and *skin surface lipids*. Skin layers are the main component of skin. Fine wrinkles, hairs and skin surface lipids are observed on the outermost surface of skin. These components exhibit very different optical behaviors according to their structures. We now give details of the physio-anatomical characteristics of these components.

2.3.1 Skin Layers

Skin layers consist of two different layers — the epidermis and the dermis (Level 2). As shown in Figure 2.5, there is a very clear wavy boundary between these two layers. The thickness of the skin layers differs quite a bit depending on gender, age, individual, body regions, etc. It has been found that males tend to have thicker skin layers than females [137]. On the other hand, while several studies on the relation between age and the thickness of the skin layers have been reported [43, 63, 107], a clear relationship has not yet been found. Conditions of skin such as water retention also differ depending on region, age, and individual. For instance, skin layers of older people have less ability for water retention, since the NMF tends to decrease with age. It has also been reported that the transparency of the stratum corneum (the outermost sublayer of the epidermis) decreases with the amount of water contained in it [138].

2.3.2 Hairs

Hairs behave as strong scatterers of light and affect the surface reflection of skin. Hairs are distributed all over the surface of the body except

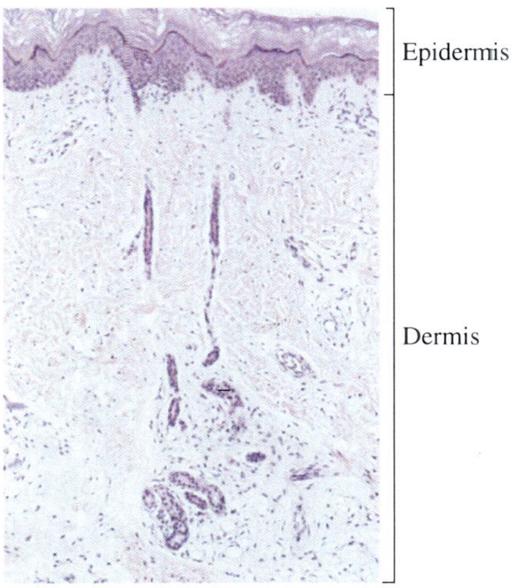

Fig. 2.5 Skin layers consist of the epidermis and the dermis. (Photo courtesy of Christopher Shea, Duke University Medical Center.)

for the palms of the hands, the soles of the feet, the tops of the feet, and mucocutaneous junctions [78, 144]. Hair grows from a hair follicle which resides in the dermal layer and opens onto the surface of skin. The bottom end of a hair follicle sits close to the boundary between the dermis and the subcutis. The color of hair is determined by the amount of eumelanin (dark brown) or pheomelanin (yellow reddish brown) it contains. Hair that does not contain melanin is white.

Hair can be categorized into two types — terminal thick hair and vellus fine hair. Terminal hair is usually hard and long (up to 3 feet long [1]). Examples include the hair on the head, eyelashes, and eyebrows. Vellus hairs, sometimes called "peachy fuzz," are soft and short (approximately 1 mm long [1]). They can be found on the cheek, forehead, arms, etc. In body regions that are covered with vellus hair, the hair plays an important role in the visual appearance of the region.

2.3.3 Skin Surface Lipid

Skin surface lipid plays an important role in surface reflection from skin. It reflects incident light specularly and makes skin appearance shinier. Skin surface lipid can be observed in most body regions, but its volume varies from one region to another. Furthermore, it also varies with respect to gender and age.

Skin surface lipid forms a thin film called the skin surface lipid film (SSLF) on skin layers. The SSLF protects excess evaporation of water from skin and keeps skin moisturized and smooth. The basis of the SSLF is sebum which is a yellowish oily liquid secreted by sebaceous glands. Sebaceous glands are present in the dermal layer and are connected to hair follicles. They cover a large region of the face, the middle of the back and the chest [144]. Although the contents of sebum vary depending on the body region, the composition of sebum is roughly estimated to be 30% of free fatty acid, 33% of triglycerides, 15% of wax, 5% of sterol esters, 5% of squalene, and 7% of paraffin [151]. After sebum is secreted, it gets mixed with sweat and lipid. This results in an emulsified film over the surface of skin [151]. The SSLF spreads over the skin surface as an unevenly distributed amorphous sheet. The thickness of the SSLF varies from one body region

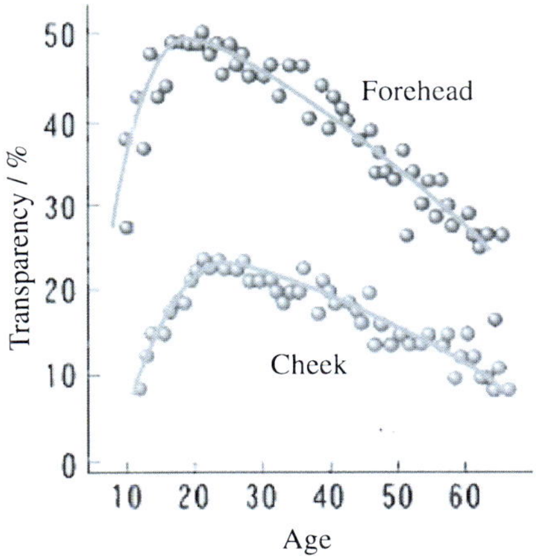

Fig. 2.6 The relation between the volume of sebum secretion and age. The volume of sebum is measured as the transparency of a thin paper used to wipe the corresponding facial area. The transparency is proportional to the volume of sebum. The graph show that the volume of sebum changes with age and peaks in the mid-twenties. (Courtesy of Kao Corporation.)

to another and is roughly estimated to range from 0.01–2.1 μm [151]. The degree of sebum secretion depends on age, gender, body regions, etc. For instance, the relation between the volume of sebum in the cheek/forearm and age is shown in Figure 2.6. Sebum secretion starts to decrease from the mid-twenties and at the age of 50 becomes similar in degree to that in the case of a 10-years old [79].

2.3.4 Fine Wrinkle

Fine wrinkles are observed on the surface of skin layers. They make the morphology of skin rough and therefore contribute to the diffusion of incident light at the surface of skin.

Fine wrinkles, also referred to as *sulcus cutis* or *glyphic patterns*, are furrows that cross each other and form squares, rectangles, triangles, and trapezoids. As shown in Figure 2.7, lines of fine wrinkles are classified into *primary lines* and *secondary lines* based on their directions, widths, and depths. The primary lines are wide and deep. They are oriented parallel to each other. The secondary lines are shallower and narrower than the primary lines. The secondary lines are diagonal to

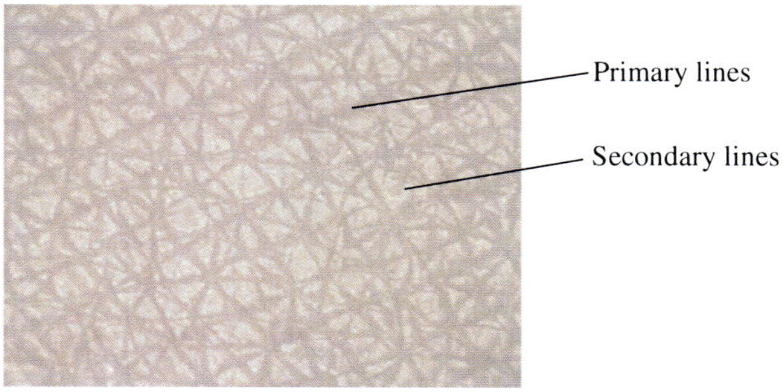

Fig. 2.7 Microscope image of fine wrinkles in the forearm. Lines are classified into *primary lines* and *secondary lines* based on their directions, widths, and depths. (Courtesy of Kao Corporation.)

the primary lines, and hence divide the skin surface into sub-regions with various shapes [184].

The morphology of these lines varies across different ages, body regions, etc. [31, 110, 140, 147]. For instance, as shown in Figure 2.8, the morphology of fine wrinkles in the volar side of the forearm is strongly related to age. As one ages, the primary lines tend to become wider and deeper, aligning with a few preferential directions. In contrast, the secondary anisotropy is unrelated to aging [184]. Fourier analysis of images of fine wrinkles has verified these results [73, 172]. These studies have revealed that fine wrinkles of young skin primarily exhibit

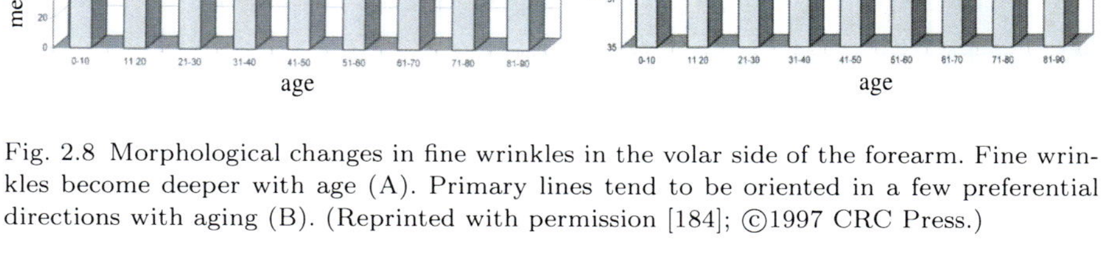

Fig. 2.8 Morphological changes in fine wrinkles in the volar side of the forearm. Fine wrinkles become deeper with age (A). Primary lines tend to be oriented in a few preferential directions with aging (B). (Reprinted with permission [184]; ©1997 CRC Press.)

high spatial frequency while those of aged skin have lower spatial frequencies.

2.4 Level 4: Skin Features

The features of skin (Level 3) constitute higher-level structure. Typical types of skin features are *wrinkles*, *pores*, *freckles*, *spots*, and *moles*. They are clearly visible to our naked eyes, and we are very familiar with their appearances.

As shown in Figure 2.9, pores and wrinkles have three-dimensional structures while freckles and spots have more or less two-dimensional color variations. Their optical properties strongly depend on their morphologies. For example, the appearance of a wrinkle depends on its geometry, including its depth and width. The appearance of freckles is determined by their colors, sizes, and spatial distribution.

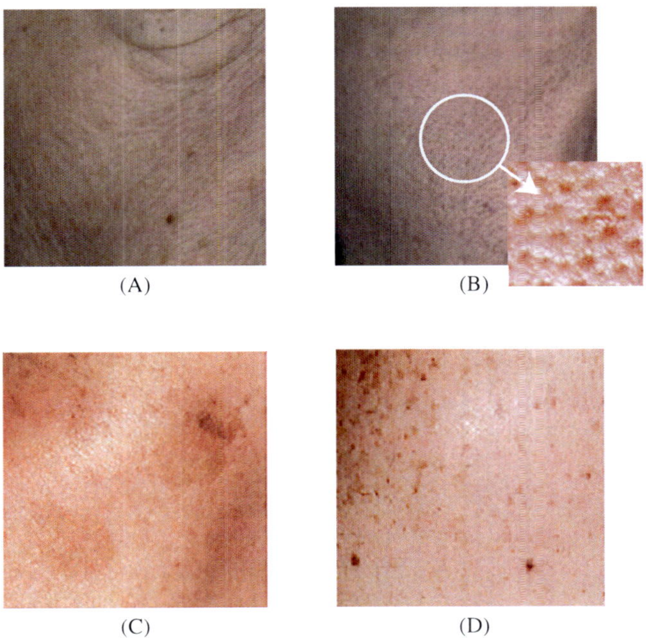

Fig. 2.9 Skin features: shallow and deep wrinkles (A), pores (B), spots (C), and freckles (D). (Courtesy of Kao Corporation.)

2.4.1 Wrinkle

As seen in Figure 2.9(A), wrinkles are folds of skin that are formed through the process of skin deformation. Wrinkles can be classified into two types: shallow wrinkles and deep wrinkles. Shallow wrinkles are usually the end result of the distortion of the epidermis caused by water-loss. On the other hand, deep wrinkles are mainly formed by the distortion of the dermis due to loss of elasticity induced by decrease of collagen and elastin fibers.

Wrinkles can be categorized in another way — expressive wrinkles and wrinkles caused by aging. Wrinkles due to aging are formed by the process mentioned above. Expressive wrinkles are particularly found on the face during expressions, irrespective of the age of the person. These wrinkles are formed by the deformation of the dermis and the subcutis as a result of muscle movement. In other words, expressive wrinkles are temporary wrinkles [186].

The morphology of wrinkles varies according to age, body regions, etc [2, 66, 90, 109, 126]. Figure 2.10 shows the relation between depth/width of wrinkles around eyes and age [2]. As one gets older,

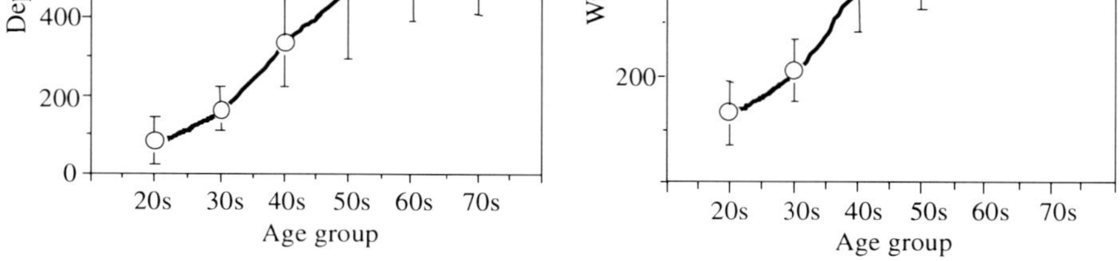

Fig. 2.10 Age-related changes in the depth (A) and width (B) of wrinkles at the eye corners. Age groups: 20s, 20–29 years; 30s, 30–39 years; 40s; 40–49 years; 50s, 50–59 years; 60s, 60–69 years; 70s, 70–80 years (total number of samples = 101). (Courtesy of G. Imokawa; reprinted Figure 6 with permission [2]; ©2002, Blackwell Publishing.)

wrinkles become deeper and wider. As we can see from these data, wrinkles are strong indicators of aging.

2.4.2 Pore

Pores (see Figure 2.9(B)) are widely dilated orifices of sebaceous and sweat glands that appear on the surface of skin. These glands are found within hair follicles. Hence, pores do not pass through skin completely and are always terminated at sweat glands at the dermis level. The sebaceous glands have lobular structures and range in size from one region to another. Many pores can be formed in sebum-rich regions such as the face, since pores are connected to sebaceous glands.

Pores act not only as the openings from which oil and sweat are secreted, but also serve as channels that carry certain substances to the deeper dermal layer. This is the reason why materials applied to the surface of skin with hair can penetrate more deeply into the skin than in the case of hairless skin [144].

2.4.3 Freckle, Spot, and Mole

As shown in Figures 2.9(C) and 2.9(D), freckles, spots, and moles are dark brown marks that can be found on the surface of skin. They have no specific shapes but usually have very clear contours. All of these features are pigmentary deposits of melanin. As explained in Section 2.1.3, melanin is usually metabolized through the process of turnover and is finally evacuated from the surface of skin. However, this metabolism sometimes does not work well because of metabolic malfunctions caused by ultraviolet light, stimulation, aging, etc. In such cases, melanin remains in the epidermis or penetrates into the dermis and forms pigmentary deposits. Ultraviolet light, stimulation, and aging sometimes cause excessive production of melanin by melanocytes. In such cases, melanin cannot be evacuated completely and tends to remain in the epidermis. This is another cause for pigmentation of skin.

The colors of these features are generally dark or yellowish brown. But, in fact, they tend to change according to their location on skin [161]. For example, the colors of spots become slightly bluish in cases when melanin is present in the dermis.

2.5 Level 5: Body Regions

The components of skin from Level 1 through 4 make up *body regions*. When we study the skin appearance of body regions, it is very important to know the shape/size as well as the physiology of each body region. Some body regions are shown in Figure 2.11. Their shapes range widely. In order to understand the morphological characteristics of these regions, it is important to know their shapes quantitatively. However, body regions are usually very complicated in shape and their shapes can vary between individuals, ages, genders, and races. As a result, it is laborious to collect quantitative data related to the shapes of body regions.

Fortunately, data measured in the field of anthropometry provide us with a good start. For instance, some data related to North American Caucasians are given in [54]. As an example, Table 2.1 [54] shows data related to the shapes of noses of North American Caucasians. One

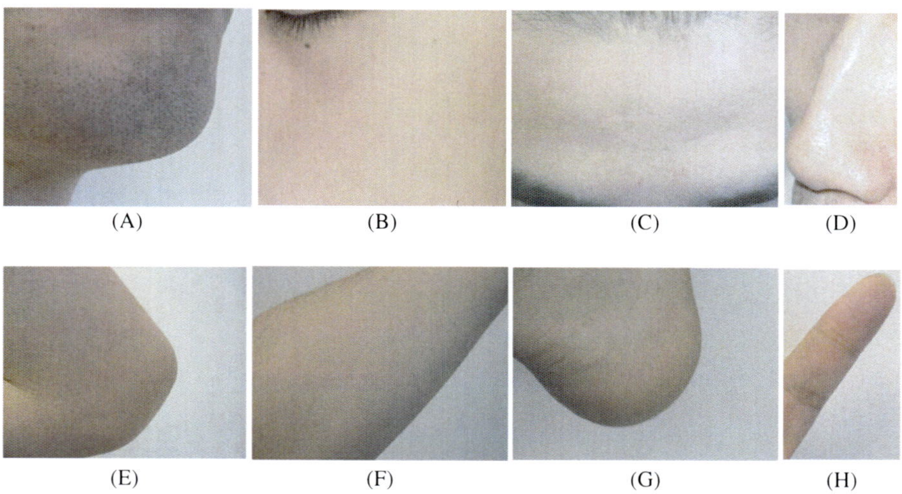

Fig. 2.11 Images of body regions: Chin (A), cheek (B), forehead (C), nose (D), elbow (E), forearm (F), heel (G), and finger (H).

Table 2.1 The nasal shape of North American Caucasians. The morphology of the nose varies according to age and gender. These data are from [54].

Age/	Width/mm		Height/mm		Bridge length/mm		Nasal tip protrusion/mm	
Years	Male	Female	Male	Female	Male	Female	Male	Female
1	26.5	25.9	30.9	29.2	27.6	25.9	10.1	10.2
5	28.9	28.5	38.9	39.3	33.5	32.8	13.3	13.1
10	30.2	29.6	45.0	44.5	40.1	39.0	16.5	16.6
15	34.2	31.7	51.9	49.2	47.4	44.8	18.6	18.7
19–25	34.9	31.4	54.8	50.6	50.0	44.7	19.5	19.7

can clearly see that nasal shapes vary depending on gender and age. For males, the width of the nose stabilizes around the age of 14. The height and bridge length of the nose stop changing around the age of 15. Nasal tip protrusion matures at the age of 16. For females, the width and height of the nose reach their full sizes at the age of 12. The nasal bridge plateaus at 13 years of age. The nasal tip protrusion fully matures around the age of 14. Usually, the nose sizes of males are larger than those of females.

The physio-anatomical properties of the skin components discussed in Section 2.1 through 2.4 tend to vary depending on the body regions. This variation also causes the optical properties of skin to vary between body regions. We now give several examples of these variations.

In cellular-level elements (Level 1), the density of melanocytes varies between different body regions. Regions that are directly exposed to sunlight (e.g., facial skin) have more melanocytes and melanin than other regions (e.g., torso).

In the skin layers (Level 2), the thickness of the epidermis varies considerably depending on the body region. Regions that are frequently subjected to friction and impacts usually tend to be thicker than regions that are free from such stimulations. For example, the eyelid is one of the thinnest regions (the epidermis is 0.04 mm and the dermis is 0.3 mm) while the sole of the foot, palm of the hand, and the back are the thickest regions (the epidermis is 1.6 mm in a sole and the dermis is 3.0 mm in the back). In particular, the stratum corneum varies considerably in thickness [145].

In skin (Level 3), the morphology of fine wrinkles differs from one body region to another [184]. Figure 2.12 shows fine wrinkles in the

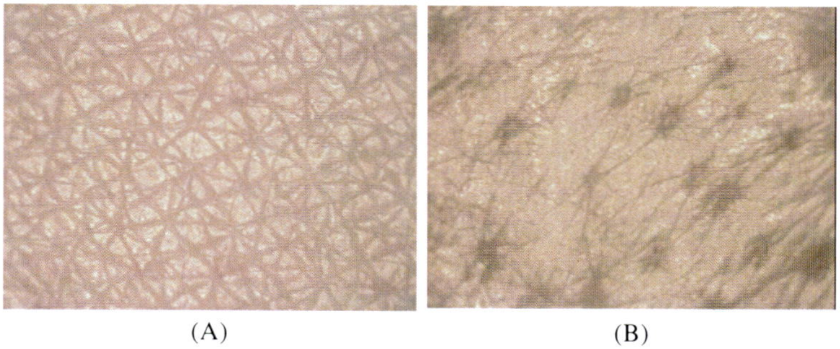

Fig. 2.12 Fine wrinkles on the forearm (A) and the cheek (B) of the same person. The morphology of fine wrinkles tends to vary between body regions. (Courtesy of Kao Corporation.)

forearm and the cheek of the same person. Fine wrinkles in the forearms are usually deep and well-ordered, whereas those in cheeks are shallow, blurry, and distorted. The volume of secreted sebum also varies within each body region. In Figure 2.13, the variation of sebum secretion over the face is shown. More sebum is secreted on the forehead, the nose, and the chin compared to other regions such as the cheeks. These sebum-rich regions are commonly referred to as the *T-zone*. The scalp and the armpits are also sebum-rich sites. In these regions, the thickness of the SSLF is more than 4 μm. On the other hand, in sebum-poor regions such as extremities, the thickness is less than 0.5 μm [151].

Fig. 2.13 Regional differences of sebum secretion over the face. The forehead, the nose, and the chin are sebum-rich areas and are commonly referred to as the *T-zone*. (Courtesy of Kao Corporation.)

With respect to skin features (Level 4), in the case of the face, there tends to be more wrinkles in the forehead, around the eyes and the mouth than other sites such as the cheeks and the chin. Besides these examples, there are several other regional differences in the physiology and anatomy of skin components.

3

Models for Human Skin Appearance

The physio-anatomical components reviewed in Chapter 2 exhibit a wide variety of optical and visual phenomena. In order to analyze and synthesize the appearance of skin, one needs computational models that accurately describe these visual phenomena. In this chapter, we will review existing computational models starting from the finest scale of the taxonomy shown in Figure 1.3. In Section 3.1, we will describe the absorption properties of chromophores and the scattering properties of a single cell and of a fiber (Level 1). Then, in Section 3.2, we will describe the scattering and absorption properties of the epidermis and the dermis (Level 2). We will also review physics-based models that represent light transport in these skin layers. In Section 3.3, we will review early studies on the optical properties of skin layers, hairs, fine wrinkles, and the SSLF (Level 3). In particular, the directional variations of their reflectance properties are described in detail. In Section 3.4, we will review early studies on skin texture appearance (Level 4). Skin texture appearance also has directional dependency. Finally in Section 3.5, we will review regional variations of the optical and visual properties reviewed in Sections 3.1–3.4.

3.1 Cellular Optics

We begin with the optical properties of cellular-level elements: *cellular optics*. Absorption in skin over the visible spectrum is mainly caused by two types of chromophores: melanin and hemoglobin. On the other hand, scattering is mainly caused by fibers, cells, and cellular organelles. Absorption and scattering at this cellular level will serve as the building blocks of light transport at the higher levels. In particular, they suggest that skin is a strong forward scattering medium at the microscale.

3.1.1 Absorption of Cellular-Level Elements

Absorption of light by skin, in the visible wavelength range, is mainly due to melanin and hemoglobin. Absorption by other cellular-level elements such as cells or fibers is negligible. As a result, the spectral reflectance property of skin as a whole can be explained by the light absorption by melanin and hemoglobin [128, 166].

The spectral absorption properties of melanin and hemoglobin have been reported in many early studies [6, 82, 141, 160, 193, 194]. For example, Figure 3.1 shows the spectral reflectance curves of melanin, oxy-hemoglobin, and deoxy-hemoglobin measured *in vitro* by Anderson and Parrish [6].

In Figure 3.1, one can see that absorption by melanin decreases steadily with wavelength, resulting in a brownish color (rich in red and poor in blue). Both oxy- and deoxy-hemoglobin have local maxima in their absorption spectra in the short and middle wavelength ranges of the visible spectrum. The absorption spectrum of oxy-hemoglobin has two maxima ("W" pattern) around 550 nm. In contrast, the spectrum of deoxy-hemoglobin has a single peak around 550 nm.

3.1.2 Scattering from Cellular-Level Elements

Scattering properties of cellular-level elements greatly depend on their refractive indices. Table 3.1 shows refractive indices of subcellular elements such as cytoplasm, mitochondria, and collagen fiber [49]. Since the refractive index of a collagen fiber in skin has not been reported,

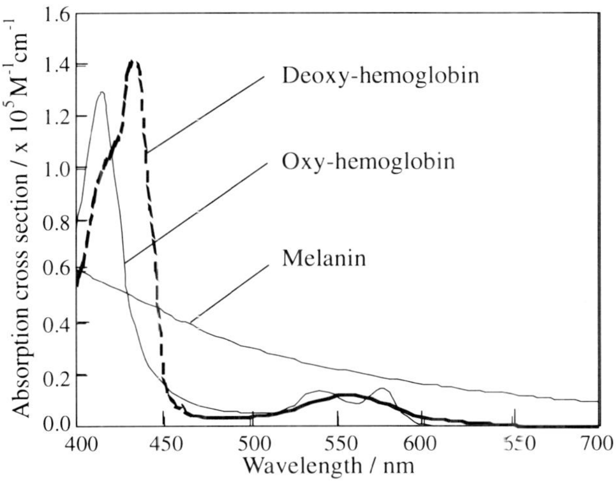

Fig. 3.1 Spectral absorption of melanin, oxy-hemoglobin and deoxy-hemoglobin [6]. Absorption by melanin increases steadily toward shorter wavelengths in the visible spectrum. Both oxy- and deoxy-hemoglobin have local maxima around 430 and 550 nm, respectively. Oxy-hemoglobin has a characteristic "W" pattern around 550 nm.

that of tendon is shown in Table 3.1. The refractive index of collagen fibers vary appreciably due to birefringence [85, 117, 130, 146, 188].

Figures 3.2(a) and 3.2(b) show schematics of scattering due to a cell and a fiber, respectively. These elements have higher refractive indices than the surrounding medium or air. Such mismatches of refractive

Table 3.1 Refractive indices of cellular-level elements [49].

Cellular-level elements	Refractive index	Remarks
Cytoplasm	1.38	Rat liver cell
Cytoplasm	1.35	
Cytoplasm	1.37	Hamster ovary cell
Cytoplasm	1.358 ~ 1.374	
Cortical cytoplasm	1.35 ~ 1.37	
Nucleus	1.39	
Cell membrane	1.46	
Mitochondria	1.40	Rat liver cell
Protein	1.5	
Deried protein	1.58	
Melanin	1.7	
Collagen fiber	1.32–1.45 (axis) 1.40–1.61 (radial)	Dry demineralized tendon

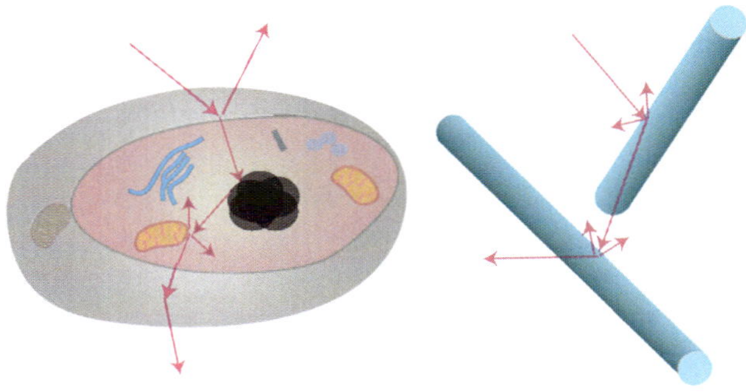
(a) (b)

Fig. 3.2 Illustration of scattering from a cell (a) and a fiber (b). Refractive index mismatches between a cytoplasmic organelle, a cell membrane or a fiber and its surroundings causes scattering.

indices result in scattering. Interestingly, the melanin granule has a much higher refractive index than other elements. As a result, a melanin granule behaves as a strong scatterer as well as an absorber [47, 170].

The scattering properties of cellular-level elements also vary with their sizes. When the size of the element is close to the wavelength of the incident light, its scattering property can be described by *Mie scattering*. Such cellular-level elements include mitochondria, nuclei, and collagen fibers [81, 89, 118]. In contrast, when the size of the element is much smaller than the wavelength, its scattering property can be described using *Rayleigh scattering*. The intensity as a result of Rayleigh scattering is proportional to λ^{-4}, where λ is the wavelength of the incident light. Cellular-level elements that exhibit Rayleigh scattering include membranes and the banded ultrastructures of collagen fibrils [81, 89, 118].

Cells are one of the main scatterers in the epidermis. The scattering property of a cell has been quantitatively investigated using different physics-based models and numerical methods such as *Mie theory* [16, 20, 25, 123, 149, 155], *Rayleigh-Gans theory* [20, 21], *anomalous diffraction approximation* [158, 159], *Fraunhofer diffraction* [21], *finite difference time domain (FDTD) method* [47, 48, 49, 50], *T-matrix method* [133], *multipole scattering* [175], and *Monte-Carlo simulation* [124].

Although the possibility that the nucleus acts as the main scatterer in a cell cannot be eliminated [16, 155], most early studies have reported that a significant portion of the scattering is due to the cell membrane, small elements in the nucleus, and the small cytoplasmic organelles such as mitochondria [19, 20, 21, 25, 47, 48, 49, 50, 122, 123, 124, 175]. These studies have also shown that the cell membrane produces scattering at larger angles. It has also been shown that a cell is basically a forward scatterer and backward scattering increases as the volume fraction of smaller organelles increases.

The main scatterer in the dermis is the collagen fiber [6, 7]. Although, the scattering property of a collagen fiber has not been measured, it has been simulated using several different physics-based models [47, 146]. For example, Drezek et al. computed the angular scattering pattern of a collagen fiber with the FDTD method [47]. In this simulation, the collagen fiber was approximated to have a cylindrical shape of 3 μm diameter and 20 μm length with a refractive index of 1.36. The results of the simulation show that a collagen fiber is a strong forward scatterer. Saidi et al. also simulated the angular scattering distribution of a collagen fiber with cylindrical Mie theory and concluded that the scattering of a collagen fiber is very strong in the forward direction [146]. These studies clearly suggest that skin is a strong forward scattering medium at the microscale.

3.2 Cutaneous Optics

Scattering properties of skin layers are not necessarily the same as those of cellular-level elements. For example, the dermis is composed of many collagen fibers that are densely packed. As a result, scattering by the dermis can be described as multiple scattering by these collagen fibers. If we could render every single optical event due to each cellular-level element in the skin layers, we could estimate the optical properties of the skin layers accurately. However, it is prohibitively expensive to render all these optical events because of the large number of cellular-level elements and their inhomogeneous distributions within the skin layers. Instead, we can estimate the optical properties of skin layers by assuming them to be optically homogeneous media. Previous works on the optical properties of skin layers or *cutaneous optics* [7] have employed this approximation.

In this section, we will review these previous results. We will first describe in Section 3.2.1 how light propagates in the skin layers to understand the optical characteristics of each skin layer in isolation. Then, the optical parameters that represent scattering and absorption properties of skin layers will be introduced in Section 3.2.2. Finally, in Section 3.2.3, we will review several physics-based models that describe light transport in skin.

3.2.1 Optical Pathways in Skin

Light incident on skin undergoes very different pathways in each of the skin layers. Figure 3.3 shows a schematic of the optical pathways in the skin layers. We will use this schematic to describe the light

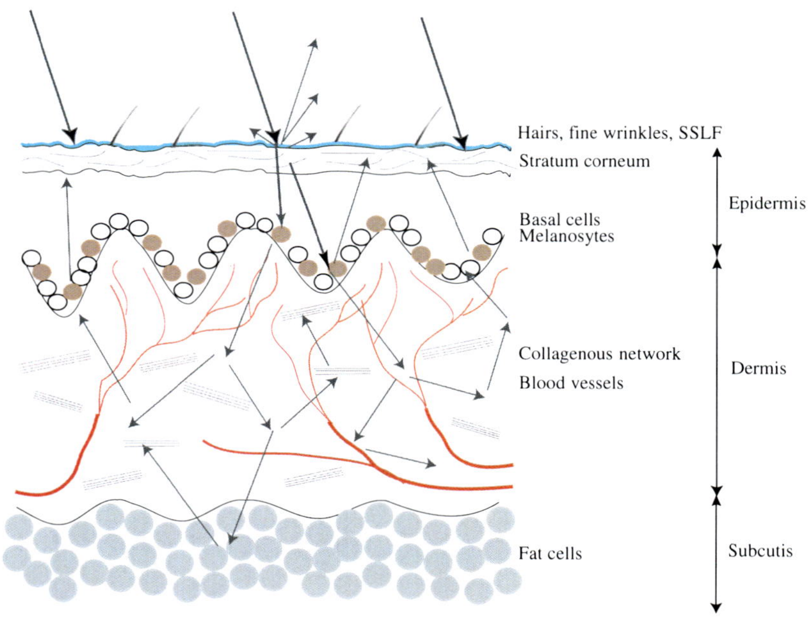

Fig. 3.3 A schematic of the optical pathways in skin. Part of the incident light is reflected at the surface of skin. The rest of the light penetrates into the skin layers. In the epidermal layer, the light rarely experiences multiple scattering but is absorbed by melanin. In dermal layer, the light is scattered multiple times by collagen fibers and absorbed by hemoglobin.

transport. In the following discussion, we will assume that (a) absorption and scattering centers are uniformly distributed within each skin layer [4, 6, 7, 103, 173], and (b) anisotropy of the light propagation caused by preferential alignments of collagen fibers [131] is negligible.

A part of the incident light is first reflected at the surface of skin due to the difference in the refractive indices of stratum corneum (approximately 1.55 [148]) and air (approximately 1.0). This reflection is called *regular reflection.* At perpendicular (normal) incidence, it represents approximately 4%–7% of the intensity of the incident light over the entire spectrum of visible light and it preserves the spectral characteristics of the incident light. The angular dependence of this regular reflection can be roughly described with the Fresnel reflection equations. Therefore, the intensity of regular reflection increases as the incident angle increases. However, it must be mentioned that regular reflection does not precisely follow the Fresnel reflection equations since the surface of skin is not perfectly smooth. In other words, the roughness of the skin surface[1] causes angular dependence of regular reflection and transmission of the incident light [6, 7, 112]. Furthermore, hair follicles, hairs, and the SSLF present on the surface of skin also cause diffusion of the regular reflection and transmission [103].

The remainder of the incident light that is not reflected at the skin surface (approximately 93%–96% of the incident light for normal incidence) enters the epidermal layer. Since the epidermis is mainly composed of well-separated scatterers (e.g., cells) [105], multiple scattering is negligible in this layer. Hence, scattering in the epidermal layer is mainly in the forward direction, reflecting the scattering property of a single cell which can be described with Mie scattering. The epidermis also contains densely packed keratin fibers that could behave as strong scatterers. However, their contribution is again negligible since the epidermis is very thin [89]. Consequently, the epidermis is a more or less transparent layer that does not produce strong enough scattering to affect the appearance of skin [6]. On the other hand, significant

[1] The average vertical peak-to-valley distance is about $0.3\,\mu m$ over a lateral distance of $20\,\mu m$ along the skin surface [184].

absorption occurs in the epidermal layer. The degree of absorption strongly depends on the volume of melanin which gives rise to the wide range of skin colors. Since the epidermal layer acts as a strong absorber and not as a scatterer, this layer is sometimes referred to as the *melanin layer*.

Figure 2.5 shows the wave-like interface between the epidermis and the dermis, which is referred to as the *dermo-epidermal junction*. Several studies have shown that this rough interface causes diffusion of light [58, 67, 112]. It has been suggested that light diffusion due to this interface increases with the change in refractive index at the interface [112].

Dermis is an optically thick and turbid medium, where a significant amount of scattering of incident light takes place [6, 7, 47, 81, 89, 146, 173]. Multiple scattering takes place in the vast network of collagen fibers (especially in the deep areas of the dermis). Therefore, scattering by the dermal layer appears to be isotropic although scattering due to a single collagen fiber is strong in the forward direction.

The scattering properties of the dermis depend on the wavelength of light. Table 3.2 shows the relationship between the wavelength and the optical path length in Caucasian skin [7]. Light scattering in the dermis varies inversely with wavelength [59, 69]. Hence, light with longer wavelength can penetrate deeper into the skin.

The light traveling in the dermal layer is absorbed by hemoglobin in the capillaries and the upper superficial arteriole and venule, the small branches of the artery and vein. Because of these scattering and absorption effects, the dermal layer can be described as a combination of two layers: a *multiple scattering layer* and a *hemoglobin layer*.

Table 3.2 The relationship between depth of penetration of incident light and wavelength in the case of Caucasian skin [7]. As the wavelength becomes longer, the incident light reaches deeper into the skin.

Wavelength/nm	Depth/μm
400	90
450	150
500	230
600	550
700	750

3.2.2 Optical Parameters for Scattering and Absorption

The scattering and absorption properties of the skin layers must be represented quantitatively in order to model light transport in skin. Absorption and scattering are usually characterized by the absorption coefficient μ_a, the scattering coefficient μ_s, and the scattering phase function $p(\hat{s},\hat{s}')$. The phase function is a probability density function that gives the probability of single scattering of light from direction $\hat{s}$ into direction $\hat{s}'$. The scattering and the absorption coefficients represent the amounts of attenuation caused by scattering and absorption, respectively. The reduced scattering coefficient $\mu_{s'}$, which is defined as $(1 - g)\mu_s$, is sometimes used to express multiple scattering instead of μ_s. Here, the factor g represents the degree of anisotropy in the scattering. This parameter is defined as the mean of the cosine of the scattering angles. Perfect forward scattering corresponds to $g = 1$ and isotropic scattering corresponds to $g = 0$. The above scattering and absorption parameters measured *in vitro* and *in vivo* are shown in Tables 3.3 and 3.4, respectively.

Table 3.3 Scattering and absorption parameters of skin over the spectrum of visible light measured *in vitro* [118].

Description	Wavelength/nm	μ_s/cm^{-1}	$\mu_{s'}/\mathrm{cm}^{-1}$	μ_a/cm^{-1}	g
Stratum corneum	400	2000		230	0.9
	415	800		66	0.74
	488	600		50	0.76
Epidermis	514	600		44	0.77
	585	470		36	0.79
	633	450		35	0.8
	415	320		4.7	0.74
	488	250		3.5	0.76
	514	250		3	0.77
	585	196		3	0.79
Dermis	633	187.5		2.7	0.8
	633		11.64	< 10	0.97
	633		23.8	2.7	
	700		21.3	1.9	
Dermis (leg)	635	244	78	1.8	0.68
Skin and underlying tissues (leg)	633	70.7	11.4	3.1	0.8
Caucasian male skin	500		50	5.1	
Caucasian female skin	500		23.9	5.2	
Hipanic male skin	500		24.2	3.8	
Forearm: fat	500		12	0.076	

Table 3.4 Scattering and absorption parameters of skin over the spectrum of visible light measured *in vivo* [118].

Description	Wavelength/nm	$\mu_{s'}/\mathrm{cm}^{-1}$	μ_a/cm^{-1}	g
Epidermis (forearm)	633	17.5	8	
Dermis (forearm)	633	17.5	0.15	
Dermis	660	$9 \sim 14.5$	$0.07 \sim 0.2$	
	633	32	0.62	
	700	28.7	0.38	
Skin ($0 \sim 1$ mm)	633	16.2	0.67	
Skin ($1 \sim 2$ mm)	633	12	0.026	
Skin (>2 mm)	633	5.3	0.96	
	633	9.08	0.17	0.05
Arm	660	8.68	0.128	0.05
	700	8.14	0.09	0.05
	633	11.17	0.072	0.09
Foot sole	660	10.45	0.053	0.09
	700	9.52	0.037	0.08
	633	16.72	0.09	0.09
Forehead	660	16.16	0.052	0.08
	700	15.38	0.024	0.06

As Graaff et al. pointed out, data measured *in vitro* tend to have larger values than those measured *in vivo* [65]. This difference can stem from differences in the sample conditions and the measurement methods used. Tables 3.3 and 3.4 clearly show that the scattering and absorption coefficients depend on wavelength. As shown in Figure 3.4, the wavelength dependency of the reduced scattering coefficient of the dermis can be described well as a combination of Mie and Rayleigh scattering. Rayleigh scattering dominates in the spectral range below 650 nm while Mie scattering plays a major role above 650 nm [81]. Based on this fact, Jacques developed an empirical equation that represents the reduced scattering coefficient $\mu_{s'}$ of the dermis as a function of the wavelength [89]. This equation is

$$\mu_{s'}(\lambda) = 2 \times 10^5 \lambda^{-1.5} + 2 \times 10^{12} \lambda^{-4},$$

where the first term is for Mie scattering and the second is for Rayleigh scattering (λ is in nanometers).

Jacques also derived equations for the absorption coefficients of the epidermis and the dermis that contain melanin and hemoglobin. The absorption coefficient of the epidermis $\mu_{a,\mathrm{epi}}$ and the dermis $\mu_{a,\mathrm{der}}$ are

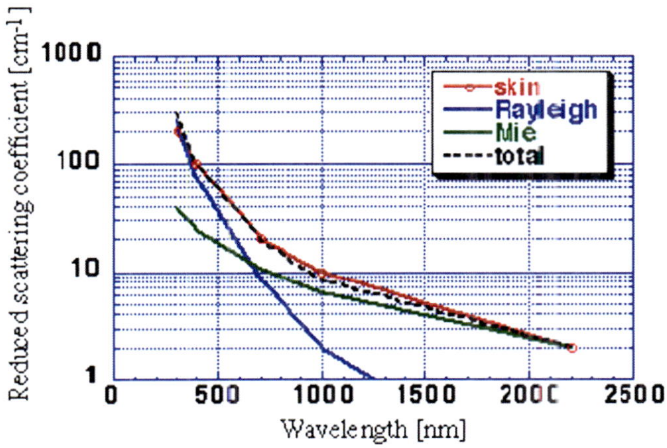

Fig. 3.4 The relationship between the reduced scattering coefficient of the dermis and wavelength. Red: tissue data, green: Mie scattering based on collagen cylinders, blue: Rayleigh limit scattering due to small-scale structure of a collagen fiber, black dashed line: sum of Mie and Rayleigh scattering, which matches the tissue data. (Courtesy of S. L. Jacques, Figure 4 in [89].)

represented as

$$\mu_{a,\mathrm{epi}} = f_{\mathrm{mel}} \cdot \mu_{a,\mathrm{mel}} + (1 - f_{\mathrm{mel}}) \cdot \mu_{a,\mathrm{skin}},$$

and

$$\mu_{a,\mathrm{der}} = f_{\mathrm{blood}} \cdot \mu_{a,\mathrm{blood}} + (1 - f_{\mathrm{blood}}) \cdot \mu_{a,\mathrm{skin}},$$

respectively, where f_{mel} is the volume fraction of melanosomes, f_{blood} is that of blood, and $\mu_{a,\mathrm{blood}}$, $\mu_{a,\mathrm{mel}}$, $\mu_{a,\mathrm{skin}}$ are the absorption coefficients of hemoglobin, single melanosome, and skin layer without any chromophores, respectively. The absorption coefficients of melanosome and the skin layer can be approximated as

$$\mu_{a,\mathrm{mel}} = 6.6 \cdot 10^{11} \lambda^{-3.33},$$

$$\mu_{a,\mathrm{skin}} = 0.244 + 85.3 \exp\left(-\frac{(\lambda - 154)}{66.2}\right).$$

Gemert et al. have proposed an empirical equation that relates the anisotropy factor g to the wavelength. Anisotropy factors of both the epidermis g_{epi} and the dermis g_{der} are given by

$$g_{\text{epi}} \sim g_{\text{der}} \sim 0.62 + 0.29 \times 10^{-3}\lambda,$$

where λ is in nm [173].

In order to determine the phase functions of the epidermis and the dermis, the angular distributions of scattering from each of them were measured *in vitro* [24, 85, 187, 189]. These studies have clearly shown the difference of the angular scattering patterns between the epidermis and the dermis. The phase function of the epidermis has been shown to be strong in the forward direction [24, 187, 189]. In contrast, the phase function of the dermis depends on its thickness; a thin sample of the dermis produces a forward-directed scattering pattern while a thicker sample greatly attenuates the on-axis intensity [85]. The difference of the phase functions between the epidermis and the dermis can be explained by considering the thickness of each of the skin layers and the scattering properties of a cell and a collagen fiber. The epidermis is very thin and is mainly composed of cells that exhibit strong forward scattering. Furthermore, cells are packed sparsely in the epidermis. Therefore, the incident light experiences forward scattering only a few times. Consequently, scattering in the epidermis is strong in the forward direction. On the other hand, the dermis is very thick and is composed of collagen fibers that are densely packed. Although a collagen fiber exhibits strong forward scattering, incident light gets scattered multiple times and hence the angular distribution of the scattering by the dermis becomes more isotropic than in the case of the epidermis.

The Henyey–Greenstein function [74] well-approximates the phase functions of the dermis [85] and the epidermis [189]. The Henyey–Greenstein function is an approximation for spherical Mie scattering [163]:

$$I(\theta) \sim \frac{1 - g_{HG}^2}{(1 + g_{HG}^2 - 2g_{HG}\cos(\theta))^{3/2}}.$$

The phase function of the epidermis is given by this equation with $g_{HG} = 0.85$–0.91 [189]. In the case of the dermis, a modified Henyey–Greenstein function fits better [85, 113]. This modified function is the combination of the two terms, one for isotropic scattering and the other for the Henyey–Greenstein function:

$$I(\theta) \sim b + (1 - b)\frac{1 - g_{HG}^2}{(1 + g_{HG}^2 - 2g_{HG}\cos(\theta))^{3/2}},$$

where b is the fraction of the isotropic scattering. The values of g_{HG} and b depend on the sample thickness. For instance, $b = 0.16$ and $g_{HG} = 0.92$ for a thin sample (20 μm in thickness), and $b = 0.58$ and $g_{HG} = 0.82$ for a thicker sample (280 μm in thickness). The first term of the modified Henyey–Greenstein function is derived from Rayleigh scattering since this scattering exhibits almost isotropic angular distribution. Hence, the modified Henyey–Greenstein suggests that scattering by the dermis includes not only Mie scattering but also Rayleigh scattering. This is supported by the fact that the reduced scattering coefficient of the dermis can be represented as a combination of two components each of which is related to Mie and Rayleigh scattering as described in Section 3.2.2. The contribution of Rayleigh scattering indicates that there must be some scatterers that are smaller than the visible wavelength of light in the dermis. Such small structures have not been clearly identified as yet. However, collagen fiber bundles are composed of fibrils that produce a striated appearance under an electron microscope. Jacques has suggested that this small-scale structure gives rise to the Rayleigh scattering of the dermis [81].

3.2.3 Models for Light Transport in Skin

Some physics-based models that can quantitatively describe light transport in tissues have been proposed in the past. These models usually require the values of optical parameters such as the scattering and absorption coefficients. Using the parameters shown in Tables 3.3 and 3.4 in Section 3.2.2, we can predict light transport in skin.

When we model light transport in an optical medium where no scattering, or very little scattering, happens (i.e., absorption is much greater than scattering), the incident light goes straight into the medium and gets attenuated exponentially due to absorption. In this case, the light transport is computed by the Lambert–Beer law [6, 7, 173, 181]. This law can be written as

$$L = (1 - R_F)E\exp(-\mu_t d),$$

where L is the radiance, E is incident irradiance, R_F is the coefficient of Fresnel reflectance for normal light incidence, and d is the thickness of the optical medium. Gemert et al. reported that the Lambert–Beer law works well only when $\mu_a > 10\,\mu_s$ [173]. Skin illuminated with visible light does not satisfy this condition since cellular-level elements act as strong scatterers for visible and near-infrared light. In other words, this condition holds only for ultraviolet and far-infrared light. Hence, the Lambert–Beer law is invalid for modeling light transport in skin.

However, a modified Lambert–Beer model can predict light transport in a highly scattering medium. The optical path length of light in a highly scattering medium is not exactly the same as the thickness of the medium since the light changes its direction as it gets scattered. Furthermore, the optical path length varies, as shown in Table 3.2, according to the wavelength of the incident light. In order to employ the Lambert–Beer law for a highly scattering medium, it should be modified to take into account this change in the optical path length. The modified law is derived by replacing the sample thickness d in the conventional Lambert–Beer law with the mean path length of the scattered light $d(\lambda)$:

$$L(\lambda) = (1 - R_F)E(\lambda)\exp[-\mu_a d(\lambda)].$$

This law was first applied to infrared light transport in the brain [41, 77, 135, 136]. Recently, some researchers have applied it to visible light transport in skin [72, 99, 152, 168]. In these studies, it was used to predict skin color. However, it does not take into account the phase function of scattering. Therefore, it is hard to use it to describe the spatial properties of light transport in skin due to scattering.

Monte-Carlo simulation is a powerful method for computing light transport in tissues [22, 57, 64, 65, 92, 93, 95, 106, 114, 131, 134, 153, 164, 171, 177, 178, 179, 182, 185]. This approach has been employed for simulating light transport in skin which has a complicated multi-layered structure. This method statistically computes the optical pathway of each photon, repeatedly. In order to obtain an accurate result, one needs to iterate the computation of photon migration many times (see for instance [178]). Hence, it is computationally very expensive.

Consequently, analytical methods are preferred in many applications, albeit with the cost of additional restricting assumptions.

The exact analytical form for light transport in a scattering medium is given by the radiative transfer equation (RTE) [132]:

$$\hat{s} \cdot \nabla L(\vec{r}, \hat{s}) = -\mu_t L(\vec{r}, \hat{s}) + \mu_s \int_{4\pi} p(\hat{s}, \hat{s}') L(\vec{r}, \hat{s}') d\omega' + q(\vec{r}, \hat{s}),$$

where $d\omega$ $[sr]$ denotes a solid angle around direction $\hat{s}$, L $[Wm^{-2}sr^{-1}]$ is the radiance, μ_t $[m^{-1}]$ is the total attenuation coefficient which is the summation of μ_a and μ_s, $p(\hat{s}, \hat{s}')$ $[sr^{-1}]$ is the (normalized) phase function of the medium and $q(\vec{r}, \hat{s})$ $[Wm^{-3}sr^{-1}]$ is the power of a source at $\vec{r}$ in the direction $\hat{s}$. The left-hand side represents the gradient of the radiance at location $\vec{r}$ in the direction $\hat{s}$. The first term on the right-hand side represents the decrease in radiance caused by absorption and scattering. The second term accounts for the increase in radiance due to the flux scattered into the direction of interest from other directions. This monochromatic equation is built upon a few assumptions: (a) skin is a homogeneous medium containing discrete scattering centers (i.e., no light interference takes place), and (b) polarization effects [5, 42, 83, 84, 87, 88, 119, 120, 157] can be neglected.

Unfortunately, a solution to this equation is not easily obtained because of its integro-differential form. However, it is possible to expand the RTE as powers of spherical harmonics (or Legendre polynomials, in which case azimuthal angles are dropped). This expansion leads to an almost exact solution to the RTE when the number of the polynomials is very large. It has been reported that a precise model for light transport in a tissue needs up to 150 spherical harmonics, which was estimated using the finite difference method [15]. Needless to say, this is still a computationally expensive model.

There are some practical models that approximate light transport in a tissue. Examples of these models are the *Kubelka–Munk* two- [6, 7, 32, 33, 171, 173, 176, 181] and four-flux theories [171], seven-flux theory [181, 192], the *adding-doubling method* [26, 142, 143, 182], *discrete ordinates* [121, 191], *path integrals* [139], and the *diffusion approximation equation* [27, 46, 55, 56, 57, 86, 94, 96, 97, 150, 180, 181, 191, 192, 195]. Among these models, the adding-doubling method and discrete ordinates are more accurate than others since they can provide solutions to the RTE with anisotropic scattering and mismatched boundaries. The adding-doubling method works well with layered media and captures the reflection from and transmission through the media, while discrete ordinates generate internal fluence. The adding-doubling method would be more useful than discrete ordinates for modeling the reflection from skin surface. Other models such as the Kubelka–Munk theory and the diffusion approximation equation are computationally faster than the adding-doubling method and discrete ordinates and therefore are widely used. These models can be derived from the RTE by assuming that the tissue is an isotropic scattering medium.

The Kubelka–Munk theory is one of the most popular and simplest approaches for computing light transport in a highly scattering medium. In such a medium, most photons experience multiple scattering before being absorbed or leaving the medium. In this case, the spatial distribution of scattered light becomes isotropic very rapidly. As shown in Figure 3.5, the Kubelka–Munk theory assumes that the radiation passing through a scattering medium can be divided into

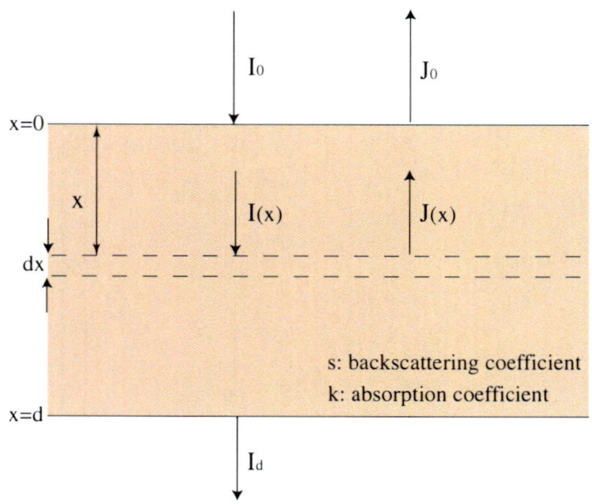

Fig. 3.5 Illustration of light transport in an optical medium modeled using the Kubelka–Munk theory. I and J depict diffuse fluxes traveling in the forward and backward directions, respectively. s and k are the backscattering and absorption coefficients of the medium, respectively. Sample thickness is denoted by d.

two "diffuse fluxes" traveling in the forward and backward directions denoted as I and J, respectively [7].

The backscattering and absorption coefficients for these diffuse fluxes (denoted as s and k, respectively) are defined as the fraction of diffuse radiation backscattered or absorbed per unit differential path length [6]:

$$dI = (-kI - sI + sJ)dx,$$
$$-dJ = (-kJ - sJ + sI)dx.$$

The reflectance ($R = \frac{J_0}{I_0}$) and transmittance ($T = \frac{I_d}{I_0}$) of the medium are related to k and s as:

$$R = \frac{(1-\beta^2)\left(\exp(Kd) - \exp(-Kd)\right)}{(1+\beta)^2\exp(Kd) - (1-\beta)^2\exp(-Kd)},$$
$$T = \frac{4\beta}{(1+\beta)^2\exp(Kd) - (1-\beta)^2\exp(-Kd)},$$

where $K = \sqrt{k(k+2s)}$, $\beta = \sqrt{\frac{k}{k+2s}}$, and d is the thickness of the medium. The assumptions made by the Kubelka–Munk theory cause some errors in certain cases. For instance, the model overestimates the radiance of scattered light in a medium of high albedo (i.e., strongly anisotropic scattering), such as the superficial area of skin, because it assumes diffuse fluxes [181]. The Kubelka–Munk theory also cannot predict the spatial distribution of light due to scattering, since the directions of scattered light are limited to only two directions.

The diffusion approximation equation is derived from the RTE under the assumption the radiance is sufficiently isotropic in the optical medium. As mentioned before, the RTE can be represented exactly with an infinite sum of Legendre polynomials. The diffusion approximation only uses the first two terms of the polynomials that represent an isotropic and a slightly forward-directed term:

$$L(\vec{r},\hat{s}) \cong \frac{1}{4\pi}\int_{4\pi} I(\vec{r},\hat{s})d\omega + \frac{3}{4\pi}\int_{4\pi} I(\vec{r},\hat{s})\hat{s}' \cdot \hat{s}d\omega.$$

With this approximation, the RTE can be rewritten as

$$D\nabla^2\Phi(\vec{r}) - \mu_a\Phi(\vec{r}) + S(\vec{r}) = 0,$$

where D is the diffusion coefficient and equals to $1/3(\mu_{s'} + \mu_a)$, $\Phi(\vec{r})$ is the fluence rate[2] of scattered light at $\vec{r}$, and $S(\vec{r}) = \int_{4\pi} q(\vec{r}) d\omega$.

Many investigators have studied the limitations and validity of the diffusion approximation for light transport in a tissue [13, 14, 17, 22, 51, 52, 53, 55, 60, 61, 62, 71, 75, 76, 98, 173, 180, 181, 190]. The diffusion approximation is valid only in a highly scattering medium where scattering dominates over absorption. Star has reported that the diffusion approximation yields the best result for an optical medium that satisfies $\frac{2\mu_a}{5\mu_s} \ll 1$ [182]. Gemert et al. have shown that it can yield "good" results when $\mu_{s'} > 10\mu_a$ and "reasonable" results when $\mu_{s'} > 5\mu_a$ [173]. Furutsu has reported that it is accurate only when $\mu_a \ll \mu_s(1 - g^2)$ [60].

In the case of skin, scattering in the dermis satisfies these conditions (recall that light is scattered multiple times in the dermis, which is primarily responsible for the scattering property of skin). The accuracy of the diffusion approximation also depends on the wavelength of incident light. Absorption by hemoglobin and melanin is greater for shorter wavelengths in the visible spectrum. Consequently, scattering is not greater than absorption in this range of wavelengths. Therefore, for skin, the diffusion approximation is not appropriate for expressing light transport of wavelengths below 440 nm [174]. It is also invalid in the "transition zone": for instance, around the interface between the optical medium and the light source. Farrell et al. have reported that the diffusion approximation is correct only under the condition $r > 1/\mu_s$, where r denotes the source-detector separation [55]. Bevilacqua et al. have shown that the diffusion approximation does not model light transport when $r\mu_s < 10$ [22].

These studies suggest that the diffusion approximation can predict light transport correctly in a region far from the light source and far from the air-tissue boundary where all photons are scattered at least several times [56, 64]. In the case of skin, this is only satisfied in the deeper level of the dermis and does not apply to the epidermis and the upper level of the dermis where scattering of radiation remains highly anisotropic. Therefore, similar to the case of the Kubelka–Munk

[2] The integral of radiance over all directions.

theory, the total fluence rate, especially in the superficial area of skin, is overestimated when the diffusion approximation is used [121]. As we will see in Section 3.3.2, this overestimation causes skin rendered using this model to appear more opaque than actual skin.

The diffuse reflectance at the surface of a scattering medium has also been modeled using the diffusion approximation equation [55, 96]. Farrell et al. have proposed a physical model that represents diffuse reflectance at a point on the surface as a function of the distance from the point where the incident light falls [55]. This equation is referred to as the *spatially resolved steady-state diffuse reflectance equation.* This can be derived from the diffusion approximation equation by using the *dipole method.*

An illustration of this method is shown in Figure 3.6. The dipole method uses two isotropic light sources that are located close to the surface of the medium to satisfy the boundary condition (the fluence rate is zero at *the extrapolated boundary*). The positive light source is located at distance z_0 beneath the surface. The negative light source is placed above the surface at distance $z_0 + 2z_b$ which is symmetri-

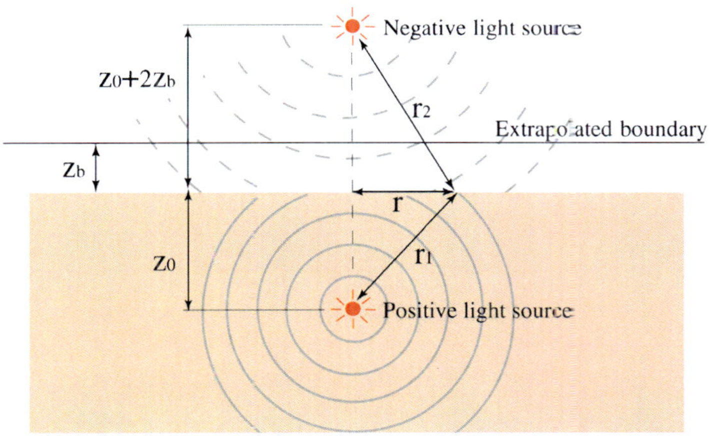

Fig. 3.6 An illustration of the *dipole method*, which is a technique for computing the fluence rate in an optical medium with diffusion approximation. This method assumes positive and negative point light sources that are located close to the boundary to satisfy certain boundary conditions. The flux from these sources determine the fluence rate in the optical medium. The spatially resolved steady-state diffuse reflectance equation can be derived from the diffusion approximation equation by using this method.

cal to the positive source with respect to the extrapolated boundary. Here, z_b is the distance from the sample surface to the extrapolated boundary. The resulting diffuse reflectance $R_d(r)$ at radial distance r is given by

$$R_d(r) = \frac{\alpha\prime}{4\pi}\left[z_0\left(\mu_{\text{eff}} + \frac{1}{r_1}\right)\frac{\exp(-\mu_{\text{eff}} r_1)}{r_1^2}\right.$$
$$\left. + (z_0 + 2z_b)\left(\mu_{\text{eff}} + \frac{1}{r_2}\right)\frac{\exp(-\mu_{\text{eff}} r_2)}{r_2^2}\right], \quad (3.1)$$

where the effective extinction coefficient is $\mu_{\text{eff}} = \sqrt{3\mu_a\mu_t'}$, the distance from the surface to the positive light source is $z_0 = \frac{1}{\mu_a + \mu_{s'}}$, $z_b = 2\frac{1+r_d}{1-r_d}D$ (r_d is the diffuse Fresnel reflectance), $r_1 = \sqrt{r^2 + z_0{}^2}$, and $r_2 = \sqrt{r^2 + (z_0 + 2z_b)^2}$. This spatially resolved diffuse reflectance equation was developed in the field of medicine to estimate scattering and absorption properties of a tissue such as skin from its diffuse reflectance. This model has also been employed for realistic rendering of skin in computer graphics to simulate *subsurface scattering* [91]. We will review this work in Section 3.3.2.

3.3 BRDF and BSSRDF

The appearance of skin varies greatly depending on the illumination and viewing conditions. This results from the directional variation of reflection by skin. In order to model skin appearance accurately, it is crucial to model the directional properties of skin reflectance. The Bidirectional Reflectance Distribution Function (BRDF) and the Bidirectional Surface Scattering Reflectance Distribution Function (BSSRDF) are appropriate functions for this purpose. In this section, we will review previous studies on the BRDF and BSSRDF of skin.

3.3.1 BRDF

The BRDF [132], f_r, is a function that relates the reflected radiance dL_r at a surface point x in the direction $\vec{\omega}$ to the irradiance dE_i due

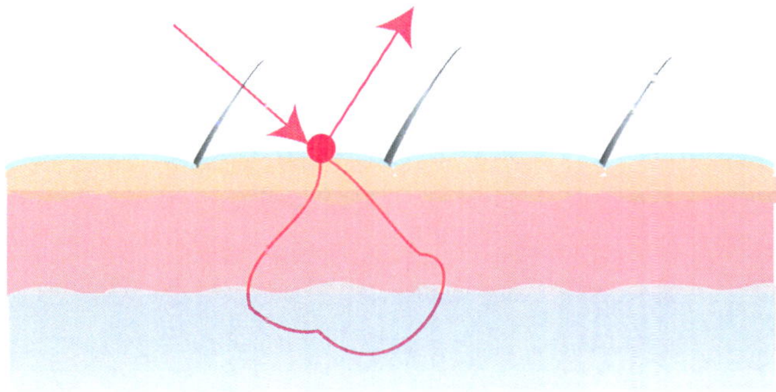

Fig. 3.7 The BRDF relates the reflected radiance to the irradiance as a function of the light source and the viewing directions. This function assumes that the point of incidence is the same as the point of reflection.

to a light source in the direction $\vec{\omega}'$:

$$f_r(x,\vec{\omega},\vec{\omega}') = \frac{dL_r(x,\vec{\omega})}{dE_i(x,\vec{\omega}')}.$$

As shown in Figure 3.7, this function assumes that the incident light hitting a surface point is reflected from the same point. The light reflected by the surface of skin tends to adhere to this assumption. Hence, the BRDF can represent the surface reflection from skin more or less accurately. On the other hand, light penetrating into the skin does not follow this assumption since it travels inside the skin layers and gets scattered, and then reemerges from surface points that are not necessarily the same as the point of incidence. Therefore, the BRDF cannot describe the body reflection (or the process of subsurface scattering) of skin. Despite this limitation, the BRDF has been widely used for rendering skin because of its low computational cost.

The BRDF of skin has been measured by several research groups. For example, the BRDF of skin measured *in vitro* is available in the Columbia-Utrecht Reflectance and Texture (CuReT) database (http://www.cs.columbia.edu/CAVE/curet/) by Dana et al. [38]. The sample skin was removed from a corpse and prepared as a flat sample. Then it was measured with 200 different combinations of incident light and viewing directions. Since the data was captured *in vitro*, the conditions of the sample such as the degree of hydration and the contents of chromophores are different from those of living skin. Such changes of physiological conditions lead to changes in refractive index, absorption and scattering properties of skin. Therefore, the measured data does not accurately represent the BRDF of living skin.

Marschner et al. have measured the reflectance of skin on the forehead over a wide range of directions *in vivo* [115, 116]. The shape of a forehead is curved. They used a CCD camera for measuring the radiometric data, a 3D range scanner for measuring the geometry of the forehead, and machine-readable targets to later register both. From the radiometric and geometric information captured by these devices, the BRDF was computed. In [115] and [116], BRDFs of four different types of skin (in terms of complexion, age, race, and gender) were measured. Figure 3.8 shows rendered images using these measured BRDFs. These images clearly exhibit natural variations in skin appearance such as shininess and tint. For example, the fourth rendered image of a South Asian (Indian) face has not only dark skin color but also a shinier appearance. Sheen usually observed around grazing angles is also seen in these images.

As can be seen in Figure 3.9 [115, 116], skin exhibits a very unusual BRDF. At small incident angles, the BRDF almost follows the Lambertian model. On the other hand, at larger incident angles, the reflectance increases significantly with the viewing angle. It is hard to describe such a BRDF using conventional physics-based models. For example, Murakami et al. have reported that neither the surface nor the body reflection of skin are correctly simulated with the *Torrance–Sparrow* [162] or the *Oren–Nayar* [129] reflection models [125]. They have also shown that Monte-Carlo simulation taking into account subsurface scattering in skin layers more accurately represents the directional variation of the body reflection component of skin. These results indicate three important points; (1) surface reflection cannot be described with conventional rough surface models such as the Torrance–Sparrow model since the assumptions such models are based on are too simple to represent the complicated geometry of real skin surface, (2) Lambert's law is inadequate for representing body reflection of skin, and (3) subsurface scattering in skin should be taken into account to describe the body reflection.

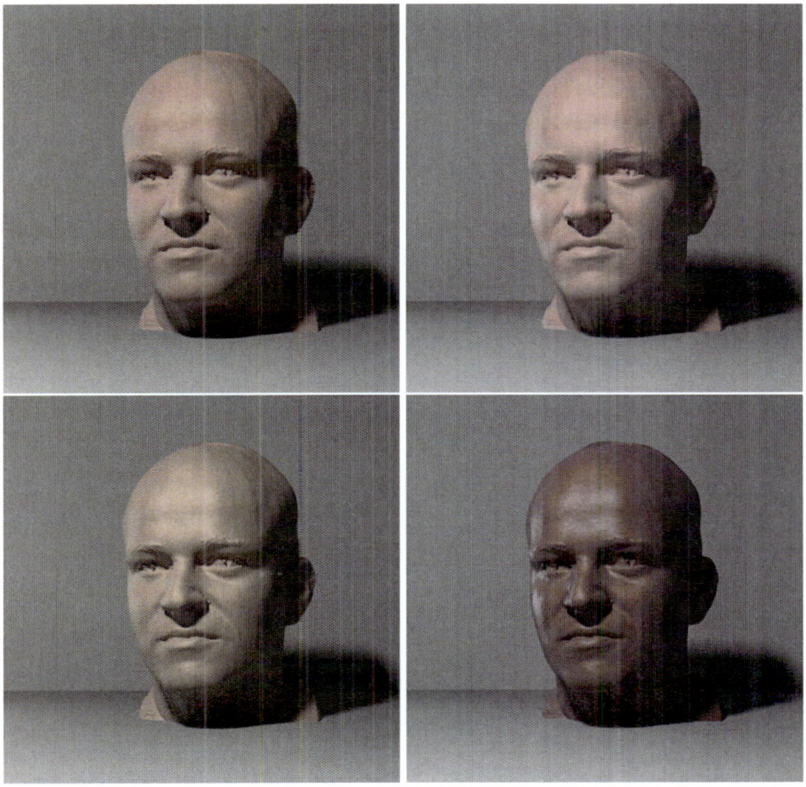

Fig. 3.8 Images rendered using BRDFs measured from four different subjects; a 43-year old Caucasian male (top left), a 26-year old Caucasian male who has a different complexion (top right), a 9-year old girl (bottom left), and a 23-year old male from India (bottom right). (Courtesy of S. R. Marschner, Figure 8 in [116].)

There are several approximate BRDF models for skin that take into account the effects of subsurface scattering [68, 156]. The first model was developed by Hanrahan and Krueger [68]. It is a linear combination of two terms, one for surface and the other for subsurface scattering. Subsurface scattering is computed with Monte-Carlo simulations, in which the scattering and absorption properties are described using the Henyey–Greenstein function. Figure 3.10 shows faces rendered using the Lambertian diffuse reflection model (the left column) and the Hanrahan–Krueger model (the middle) for 0° (the top row) and 45° (the bottom row) angles of incidence. The right column shows the

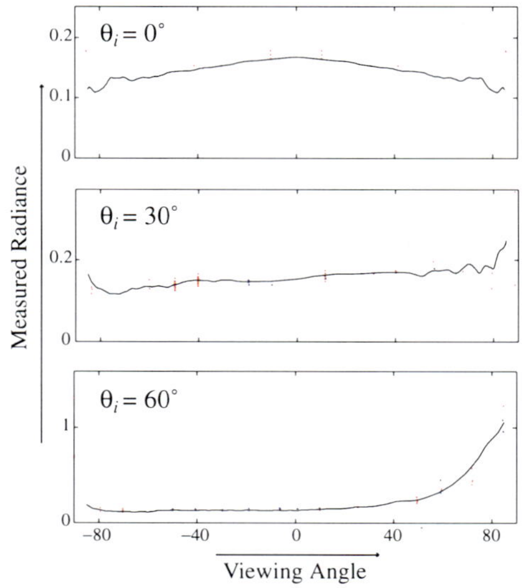

Fig. 3.9 The BRDF of typical skin (incident angle is denoted as θ_i). The BRDF of skin is almost Lambertian at small incident angles, but at larger angles the BRDF changes substantially. (Courtesy of S. R. Marschner, Figure 11 in [116].)

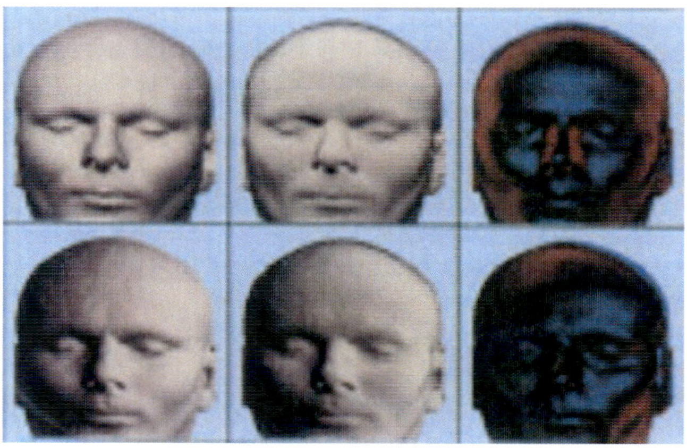

Fig. 3.10 Faces rendered with the Hanrahan–Krueger model (the middle column) and the Lambertian model (the left column) for incidence angles of 0 (top row) and 45 (bottom row) degrees. The right column shows the differences between the images produced by the two models, where red indicates more reflection by the Hanrahan–Krueger model and blue indicates the opposite. The Hanrahan–Krueger model produces a smoother "silk-like" appearance. (Courtesy of P. Hanrahan; reprinted plate 2 with permission [68]; ©1993 ACM, Inc.)

differences between the images rendered using the two models. Note that the Hanrahan–Krueger model produces a smoother and softer appearance than the Lambertian model. This result clearly shows that subsurface scattering is an important factor in reflection from skin.

While the Hanrahan–Krueger reflection model relies on statistical Monte-Carlo simulations that are computationally expensive, the model proposed by Stam is based on an analytical approach [156]. In this model, multiple anisotropic scattering in a skin layer bounded by two rough surfaces (subsurface scattering) is computed with a discrete ordinate approximation to the RTE (see Section 3.1). The reflection and refraction due to rough surfaces are computed by generalizing the Cook–Torrance reflection model [30], where the surface is treated as a Gaussian random height field. Figure 3.11 shows human heads rendered with the Stam model (right), the Hanrahan–Krueger model (center), and the Lambertian model (left). The appearance of skin rendered with the Stam model appears like a blend between the Lambertian and the Hanrahan–Krueger models.

Although there are several approximate BRDF models for skin that take subsurface scattering into account, the appearance of skin rendered with such models is still "harder" and more opaque than that of real skin. Since BRDF models assume that light is radiated by a point only if it receives incident light, the spatial contributions of radiated light due to subsurface scattering is not accounted for. In order to render more realistic skin, a BSSRDF model that includes this spatial cross-talk is

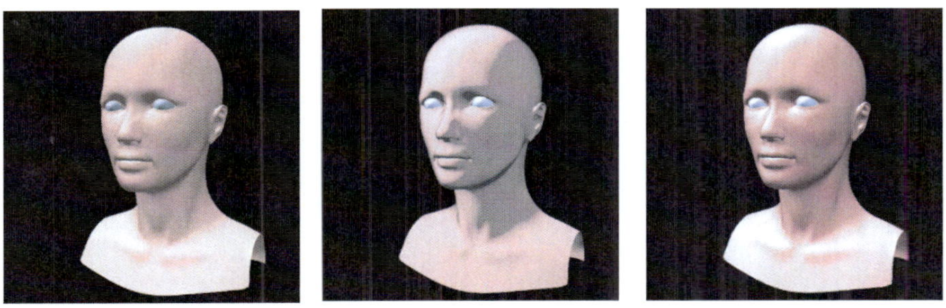

Fig. 3.11 Faces rendered with the Lambertian model (left), the Hanrahan–Krueger model (middle), and the Stam model (right). (Courtesy of J. Stam; reprinted Figure 6(a) with permission [156]; ©2001 Springer-Verlag.)

more appropriate than a traditional BRDF model. We will review early studies on the BSSRDF of skin later in Section 3.2.2.

BRDF of SSLF, Fine Wrinkles, and Hairs The skin surface lipid film (SSLF), fine wrinkles, and hairs affect the overall reflectance of skin. Furthermore, changes in the physiological conditions of these elements can significantly alter the appearance of skin.

Marschner et al. have reported that skin with a light coating of mineral oil has a more glossy appearance compared to untreated bare skin. Since the refractive index of mineral oil is close to that of the SSLF, this indicates that SSLF-rich skin looks glossy [116]. So-Ling et al. simulated the BRDF of skin covered with SSLF using the Monte-Carlo method [154]. This simulation has shown that skin with SSLF exhibits higher Fresnel reflection than skin without SSLF. The smooth air-SSLF interface results in a shinier appearance.

Obviously, the directional variation of surface reflection from skin is influenced by the morphology of fine wrinkles. As shown in Figure 3.12, reflection from skin with a dense and well-ordered network of fine wrinkles is more diffused than it is from skin with a coarse network of fine wrinkles.

It is well-known that older skin tends to be shinier than younger skin. This is due to the morphological changes in fine wrinkles (recall

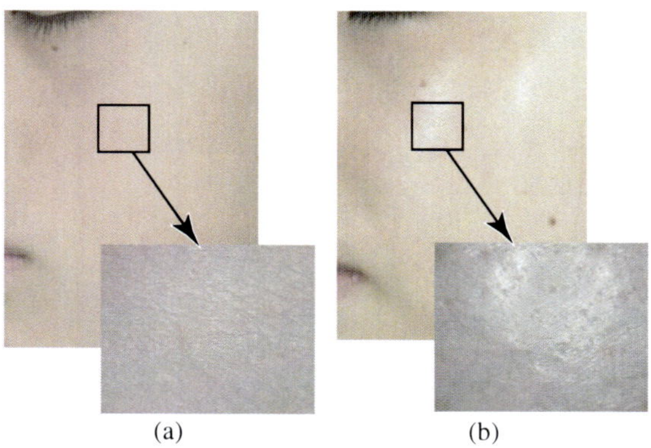

Fig. 3.12 The appearance of cheeks in which fine wrinkles form a dense network (a), and a coarse network (b). The accompanying micro-photographs of the wrinkles are taken from silicone imprints obtained from the cheeks. (Courtesy of Kao Corporation.)

that the density of fine wrinkles becomes lower with aging as shown in Figure 2.11 in Section 2.3). Such optical effects due to fine wrinkles cannot be described using conventional BRDF models. For example, as mentioned before, the Torrance–Sparrow model fails to approximate the directional variation of surface reflection from skin [125]. This is because the fine structure of skin surface does not obey the Gaussian distribution of v-grooves assumed by the Torrance–Sparrow model. In order to describe the BRDF of skin with a physics-based model, the fine geometrical structure of skin surface including the morphology of fine wrinkles should be taken into account. To our knowledge, such a model has not been developed to date.

Vellus hairs introduce a rarely acknowledged but important optical phenomenon called *asperity scattering* [100]. Asperity scattering is caused by single scattering due to point scatterers that are sparsely distributed like "cloud cover" over the surface of a material. Vellus hairs behave like good point scatterers since they are distributed sparsely over the skin surface, as shown in Figure 3.13(a).

Figure 3.13(b) shows an example of asperity scattering due to vellus hairs on the face. Hairs that sparsely cover the face brighten the

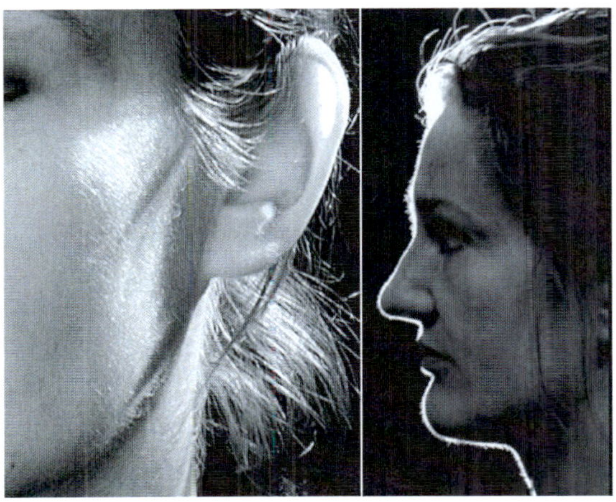

(a) (b)

Fig. 3.13 (a) Vellus hairs on the face. (b) Asperity scattering due to vellus hairs on the face. (Courtesy of J. Koenderink and S. Pont; reprinted Figures 13 and 14 with permission [100]; ©2003 Springer-Verlag.)

appearance of the face. Notice that asperity scattering is not observed at the glabrous skin of the lips.

The BRDF of asperity scattering, f_a, is represented as

$$f_a(\hat{u}, \hat{v}) \approx \frac{p(-\hat{u} \cdot \hat{v}) \Delta \mu_t}{\cos\theta_u \cos\theta_v},$$

where Δ is the thickness of a layer formed by scatterers, $\hat{u}$ and $\hat{v}$ are the directions that a photon enters and leaves, respectively, $\hat{n}$ is the local surface normal, and $p(-\hat{u} \cdot \hat{v})$ is the phase function. This equation shows that asperity scattering becomes larger at grazing angles of the viewing and illumination directions — either θ_u or $\theta_v \approx \frac{\pi}{2}$. As a result, it is clearly observed near the contours of edges and shadows in backlit conditions.

Even though the optical effect of asperity scattering is quantitatively low, it has a large influence on the visual appearance of skin. Figure 3.14 shows two disks, one with a blurred contour and the other with a sharp contour. We see that the blurred contour gives us a "softer" impression compared to the other one. Asperity scattering produces such a "soft" impression since it makes the contour of skin blurred as seen in Figure 3.13(b). Face powder cosmetic products are frequently used to give skin a "soft" appearance. The powders have scatterers that are sparsely distributed over the facial skin and hence produce asperity scattering.

Skin Color The chromatic property of skin plays a critical role in its appearance. Angelopoulou et al. have attempted to model the BRDF of skin as a function of wavelength (their study has been limited to 0°

Fig. 3.14 A "hard" and a "soft" black disk. The "soft" disk looks softer and less black than the "hard" disk. Such differences are due to the geometric attributes of the contours. (Courtesy of J. Koenderink and S. Pont; reprinted Figure 16 with permission [100]; ©2003 Springer-Verlag.)

incidence and 4° reflection angles) [8, 9, 10, 11, 12]. They developed basis functions that can represent spectral reflectance measured from various types of skin (different ages, genders, and races). A variety of basis functions were fitted to measured data and a 32-coefficient wavelet model was found to produce the best fits. Since this model is computationally expensive, a mixture of five Gaussian distributions was used as an approximation. They concluded that the Gaussian model, although less precise than the wavelet model, would be adequate for the purposes of computer graphics and vision.

3.3.2 BSSRDF

Subsurface scattering is one of the most important optical characteristics of skin. As mentioned in Section 3.2.1, several BRDF models that incorporate subsurface scattering have been developed [68, 156]. However, these models cannot describe the spatial distribution of incident light due to subsurface scattering since they are derived as BRDFs. In order to describe subsurface scattering effects more accurately, we need a reflection model in which the incoming and outgoing light locations on the surface can be different. The Bidirectional Surface Scattering Reflectance Distribution Function (BSSRDF) [132], S, is a function that relates the reflected radiance $dL_r(x, \vec{\omega})$ at a point x in direction $\vec{\omega}$ to the incoming flux $d\Phi_i(x, \vec{\omega}')$ at a point x' from direction $\vec{\omega}'$:

$$S(x, \vec{\omega}; x', \vec{\omega}') = \frac{dL_r(x, \vec{\omega})}{d\Phi_i(x, \vec{\omega}')}.$$

Figure 3.15 shows a schematic of the BSSRDF. Note that the point where the incident light hits the surface of the skin (x) does not necessarily coincide with the point where the light reemerges from the surface after traveling inside the skin (x'). This enables us to represent the spatial distribution of the reflected light due to subsurface scattering. In fact, the BRDF is a special case of the BSSRDF where light transfer inside a material can be ignored. In order to develop a BSSRDF model for skin, an expression for the light transport inside skin is needed. The light transport theories described in Section 3.1 (e.g., the diffusion approximation equation) are helpful for this purpose.

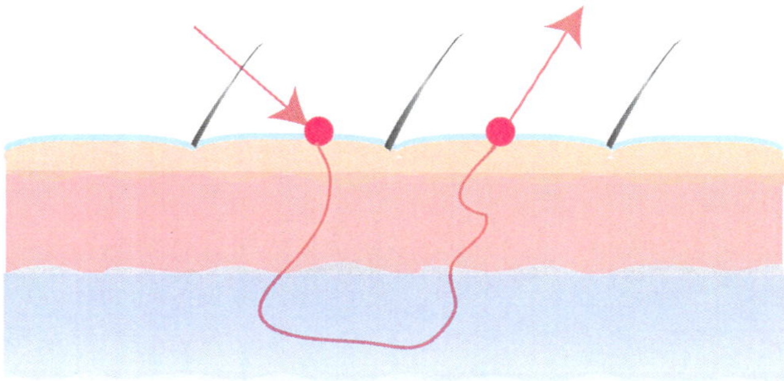

Fig. 3.15 A schematic representation of the BSSRDF. The BSSRDF relates the reflected radiance to the incoming flux as a function of illumination and viewing directions. Unlike the BRDF, the point of incidence is not necessarily coincident with the point of reflection. This enables the BSSRDF to capture subsurface scattering effects more accurately.

BSSRDF of Skin Layers There are very few studies that correctly capture subsurface scattering of skin. Debevec et al. measured the reflection of facial skin including subsurface scattering effects by capturing many images from various directions [40]. However, these measurements were not related to any physical models. The first detailed study of skin BSSRDF was done by Jensen et al. [91]. Their model is composed of a single scattering term ($S^{(1)}$) and a multiple scattering term (S_d):

$$S(x,\vec{\omega};x',\vec{\omega}') = S_d(x,\vec{\omega};x',\vec{\omega}') + S^{(1)}(x,\vec{\omega};x',\vec{\omega}').$$

In the case of skin, a significant portion of the incident light undergoes multiple scattering in the dermis. Hence, the multiple scattering term is generally larger than the single scattering term.

In order to build a physical model for the multiple scattering term, Jensen et al. employed the spatially resolved steady-state diffuse reflectance equation (see Section 3.2.3 [55]). Note that the spatially resolved diffuse reflectance equation is derived from the diffusion approximation equation. Hence, this equation represents the spatial distribution of the light due to subsurface scattering assuming isotropic radiance distribution. By taking into account the Fresnel reflection at the boundary for the incoming and outgoing light, the spatially resolved diffuse reflectance $R_d(\|x' - x\|)$ can be included in the BSSRDF model of skin as

$$S_d(x,\vec{\omega};x',\vec{\omega}') = \frac{1}{\pi} F_t(\eta,\vec{\omega}) R_d(\|x' - x\|) F_t(\eta,\vec{\omega}'),$$

where $F_t(\eta,\vec{\omega})$ and $F_t(\eta,\vec{\omega}')$ are the Fresnel transmittance in the directions $\vec{\omega}$ and $\vec{\omega}'$, respectively, for refractive index η.

As expected, the BSSRDF model yields more realistic rendered skin compared to BRDF models. Faces rendered with a BRDF and the above BSSRDF model are shown in Figures 3.16(a) and 3.16(b), respectively. The image rendered using the BSSRDF model is much softer and gives a more translucent look compared to the image rendered using a BRDF model. In the BSSRDF case, internal color bleeding caused by the absorption of light by blood in the deeper levels of skin is also observed in the shadow region under the nose. Such subtle and yet clearly visible differences between the two images demonstrate the importance of the spatial distribution of the reflected light due to subsurface scattering.

Although the appearance of Figure 3.16(b) is much more realistic compared to Figure 3.16(a), the skin surface still looks harder than real skin. This difference is mainly due to the following factors. First, the BSSRDF model developed by Jensen et al. only considers single and multiple scattering. However, light experiences scattering more

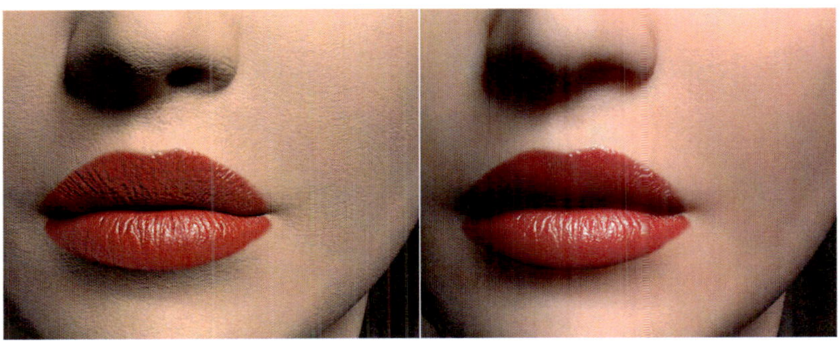

(a) (b)

Fig. 3.16 A face rendered with a BRDF model (a) and the BSSRDF model (b). The BSSRDF model produces a softer appearance. It also produces "internal color bleeding" effects which are critical to the appearance of translucent materials (see the shadow region under the nose). (Courtesy of H. W. Jensen; reprinted Figure 1[?] with permission [91]; ©2001 ACM, Inc.)

than once but just a few times in the transition zone (i.e., the upper regions of skin around the epidermal and papillary layers). Scattering in this zone is not correctly taken into account in the above model as it uses the diffusion approximation for multiple scattering. The diffusion approximation overestimates the total fluence rate in the transition zone as mentioned in Section 3.2.3. Consequently, the model does not adequately capture the transparency of the transition zone. In order to accurately describe the anisotropic light transfer in this zone, a full-blown computational simulation using Monte-Carlo methods and adding-doubling methods must be done. Second, skin was modeled as a semi-infinite single layer in the development of the BSSRDF model. However, skin is composed of several layers that have different scattering and absorption properties. It is hard to accurately describe light transport in a multi-layered structure with semi-infinite single-layered models. Third, the model uses scattering and absorption coefficients of the arm region for rendering facial skin. As we will describe in Section 3.4, the values of these coefficients vary significantly across different body regions and hence the correct coefficient values of skin must be used. Finally, the rendered skin lacks skin features such as wrinkles, freckles, and pores that are usually present in real skin. Such skin features are critical to achieving realism.

In order to render more realistic skin appearance, a more sophisticated and flexible BSSRDF model will be needed. Such a model should view skin as a multi-layered structure and include anisotropic scattering in the transition zone. Recently, Donner and Jensen [44] introduced the use of multipole diffusion theory for rendering multi-layered translucent materials. Instead of considering a single dipole to approximate the diffusion in semi-infinite homogeneous medium (see Section 3.2.3), multipole diffusion approximation uses multiple sets of dipoles to model the diffusion in a layered thin slabs of homogeneous medium. Donner and Jensen [44] combine this with Kubelka–Munk theory to compute the reflectance and transmittance profiles of multi-layered materials, which amounts to successive convolution of transmittance and reflectance profiles of each layer. In addition, Torrance–Sparrow reflection model was used to render the surface reflectance at the top layer. This multilayered model results in impressive skin renderings.

Yet, many problems remain open, such as accounting for the strong forward scattering properties of epidermis, anisotropic scattering in the transition zones, surface features including hair fossils and fine wrinkles, etc. One approach to model these complex appearance factors would be to directly reuse measured appearance (images). Weyrich et al. [183] measured the 3D face geometry and skin appearance of 149 subjects using a custom-built system consisting of a range scanner and multiple cameras and light sources covering different viewpoints and illumination directions. The measured data are then used to compute the parameter values of a skin reflectance model that models the spatially varying subsurface scattering, texture, and specular reflection. The results are used for analyzing the statistics of these parameters among different ages and gender. Although one could directly reuse measured data for realistic face rendering by interpolating the captured appearances, the excessively high dimensionality of such data renders the problem near impossible. A low-dimensional analytic model that well approximates such appearance data is very much in need, which can provide insight to compactly characterizing the appearance space even if their parameter values were not directly used. Previous studies related to light transport theories that were reviewed in Section 3.1 may prove useful in developing such an improved analytic appearance model of skin.

Skin Color The color of skin is mainly determined by the absorption and scattering characteristics of skin. Hence, a BSSRDF model that can precisely represent the light transport process will yield an accurate expression of the chromatic properties of skin. The "internal color bleeding" that is seen in Figure 3.16(b) is an example of a complex chromatic phenomenon. Unfortunately, there are no BSSRDF models that capture in detail the chromatic properties of skin. However, some previous studies on skin color provide us with helpful insights for developing more sophisticated BSSRDF models that include various chromatic effects.

Nakai et al. have reported that the spectral reflectance of various colored skins may be derived from the absorption characteristics of melanin, hemoglobin, and carotene [128]. They applied multiple regression analysis (MRA) to the spectral reflectance of various skin colors, where the underlying variables were the spectral characteristics of melanin, blood, and carotene. They found that the spectral reflectance of skin can be correctly represented with just these three variables. Since the concentration of carotene is much less than that of melanin and hemoglobin, this study suggests that skin color may be represented using just the absorption properties of melanin and hemoglobin.

Tsumura et al. have extracted the principal chromatic components that represent the various colors of normal Japanese skin using independent component analysis (ICA) [167, 168, 169]. As depicted in Figure 3.17, they showed that all skin colors lie on a plane spanned by two independent vectors denoted as $\mathbf{c}(1)$ and $\mathbf{c}(2)$ in the optical density domain. These two vectors represent the absorptions by hemoglobin and melanin, respectively [166]. As mentioned in Chapter 2, the dominant chromophores in normal skin are melanin and hemoglobin, which are physiologically independent: they are located separately in the epidermal and dermal layers. The findings of Tsumura et al. are consistent with this physiological fact. By using the approach of Tsumura et al.,

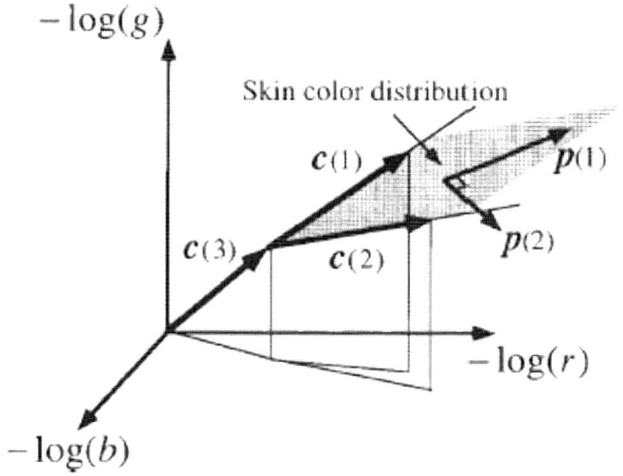

Fig. 3.17 Skin color model in the optical density domain of the tri-stimulus space. All skin colors are distributed on a plane spanned by two independent vectors that correspond to melanin ($\mathbf{c}(1)$) and hemoglobin ($\mathbf{c}(2)$), respectively. (Courtesy of N. Tsumura; reprinted Figure 4 with permission [167]; ©1999 Optical Society of America.)

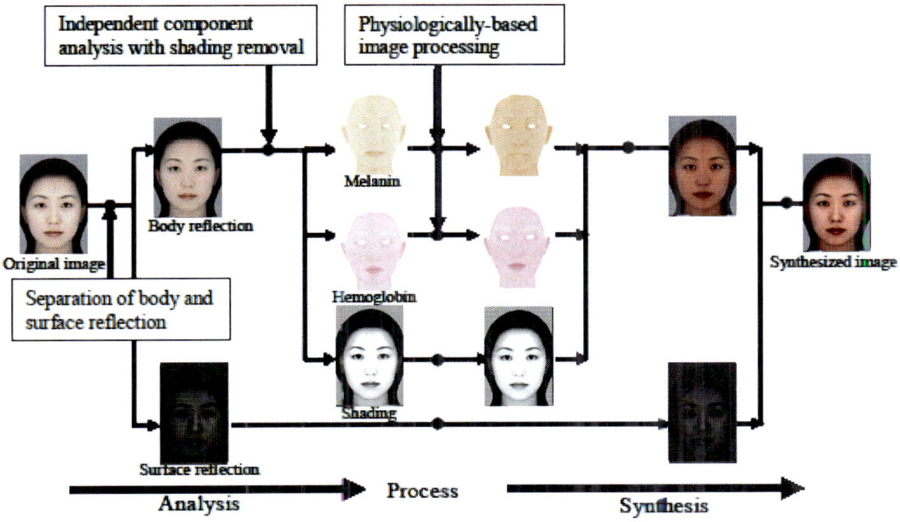

Fig. 3.18 Flow diagram for image-based skin color and texture synthesis/analysis proposed by N. Tsumura et al. (Courtesy of N. Tsumura; reprinted Figure 1 with permission [168]; ©2003 ACM, Inc. The face images in the original illustration in [168] were substituted with new face images by the above authors.)

it is possible to synthesize various skin colors [168]. Figure 3.18 shows the flow diagram for synthesizing various skin colors from an image.

The original image is first separated into images of surface and body reflection. Then, the image of body reflection is decomposed into the melanin and hemoglobin components using ICA. In this stage, shading is also removed using an inverse lighting technique. The amount of melanin and hemoglobin is changed spatially based on physiological knowledge or a database that captures the variation of skin color and texture (this process is called *physiologically based image processing*). Finally, a synthesized image is obtained by combining the processed images for body reflection and the image of surface reflection. Figure 3.19 shows synthesized images (the original image is at the center). Skin color is changed from reddish (hemoglobin-rich) to yellowish (melanin-rich) or light to dark, and these changes appear natural.

The chromatic components of skin have also been investigated based on the Kubelka–Munk two-flux theory [29, 33]. In these studies, skin is modeled as a multi-layered structure consisting of stratum

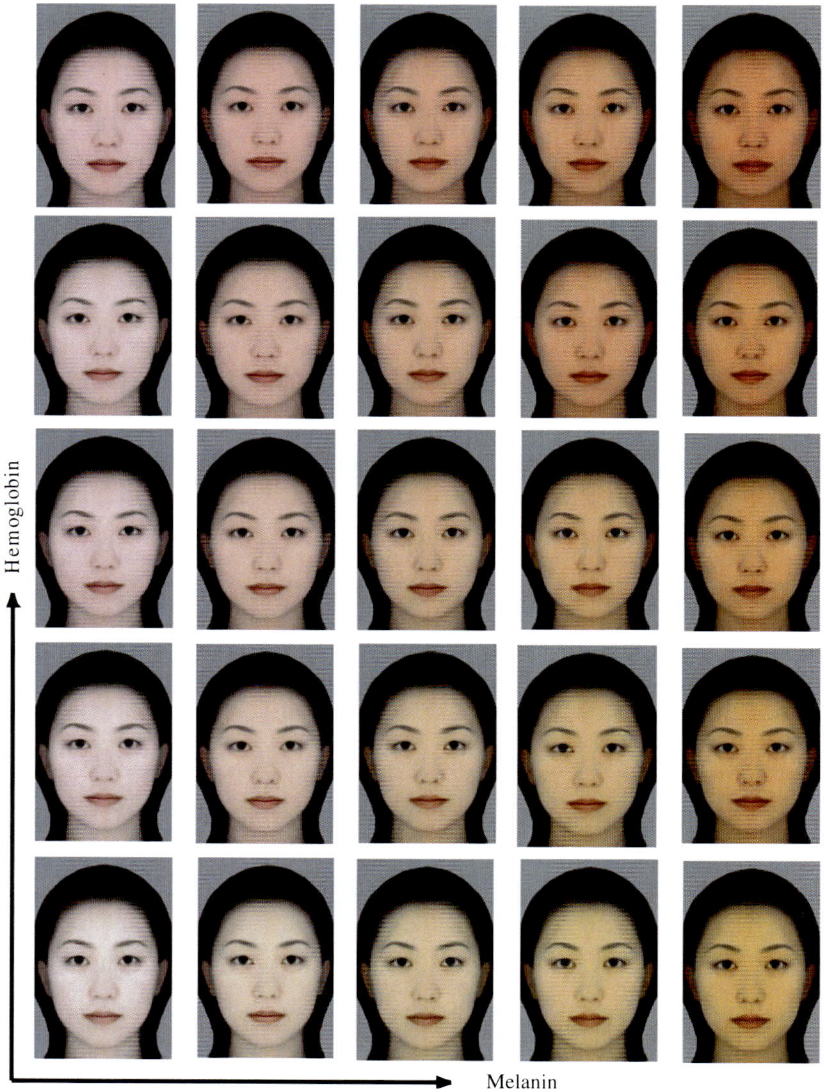

Fig. 3.19 Skin color synthesis by changing the volume of melanin and hemoglobin using the method shown in Figure 3.18. The center image is the original one. (Courtesy of N. Tsumura; reprinted Figure 7 with permission [168]; ©2003 ACM, Inc. The face images in the original illustration in [168] were substituted with new face images by the above authors.)

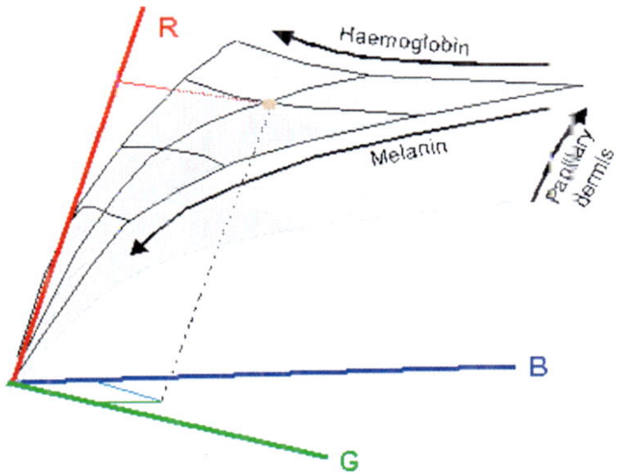

Fig. 3.20 A schematic of the relationship between two reference systems for skin color; a color system with axes representing R, G, and B; and a histological parameter system, with axes representing hemoglobin, melanin, and papillary dermis. All skin colors lie on the curved surface bounded by the histological axes. Compare this representation to the one in Figure 3.17. (Courtesy of E. Claridge; reprinted Figure 2 with permission [29]; ©2002 Springer-Verlag.)

corneum, epidermis, papillary dermis, and reticular dermis. By computing light transport through this multi-layered structure using the Kubelka–Munk theory, it was shown that all normal skin colors lie on a simple curved surface in a tri-stimulus color space (e.g., LMS [33] and RGB [29]), as shown in Figure 3.20. This curved surface is bounded by three axes corresponding to melanin, hemoglobin, and the thickness of the papillary dermis. Even though the Kubelka–Munk theory is too simple to describe light transport in skin, these studies show that normal skin color can be represented with three parameters: the absorptions by melanin and hemoglobin, and scattering by the dermis.

Physics-based models for light transport such as the diffusion approximation equation were originally proposed as monochromatic functions. However, these monochromatic functions can be extended to wavelength-dependent forms by replacing μ_a and $\mu_{s'}$ which are usually the variables of these models, with $\mu_a(\lambda)$ and $\mu_{s'}(\lambda)$. Since the chromatic properties of skin are mainly due to the absorption by melanin and hemoglobin and scattering by the dermis [29, 33, 167, 168], $\mu_a(\lambda)$ and $\mu_{s'}(\lambda)$ correspond to these absorption and scattering properties, respectively. The absorption coefficient can be represented as the sum of the absorption coefficients of melanin and hemoglobin, i.e., $\mu_a(\lambda) = \mu_{a,\mathrm{melanin}}(\lambda) + \mu_{a,\mathrm{hemoglobin}}(\lambda)$. These coefficients can be obtained from the spectral absorptions by melanin and hemoglobin that have already been measured [6, 82, 141, 160, 194] (see Section 3.1.1). Scattering by the dermis is approximated by the sum of Mie and Rayleigh scattering due to collagen fibers as mentioned in Section 3.2.2 [81]. An empirical expression for the scattering coefficients of the dermis as a function of wavelength has also been proposed (also see Section 3.2.2) [89].

Recently Krishnaswamy and Baranoski [104] used a biologically sound spectral model for simulating the appearance of skin. The model characterizes the reflection and transmission at skin layer interfaces as well as the absorption and surface/subsurface scattering properties within each skin layer. The skin appearance is then rendered using Monte-Carlo simulation using this model. Unfortunately, such a model results in a notoriously complex analytical form, in their case with 22 free parameters, which becomes difficult to handle. Donner and Jensen [45] introduced a lower-dimensional parametric spectral skin reflection model based on two-layer multipole diffusion approximation with four free parameters that controls the amount of melanin, hemoglobin, and the oiliness of the skin. By building upon these work and incorporating other spectral effects as well as spatial variation of reflectance due to the inherent inhomogeneity of skin layer components as well as facial features, one may be able to develop a wavelength-dependent BSSRDF model that captures in greater detail the chromatic phenomena that are demonstrated by skin.

3.4 BTF

The appearance of skin is noticeably uneven in the mesoscale since skin has various structural details (e.g., pores and wrinkles) and color variations (e.g., spots and freckles). A model of skin appearance must therefore include such features, often referred to as *skin texture*. For instance, skin with a lot of wrinkles and spots looks older. The rendered faces shown in Section 3.3 look unrealistic even though the BRDF or the BSSRDF models used are close to that of real skin. This is partly due to the lack of skin features in these renderings.

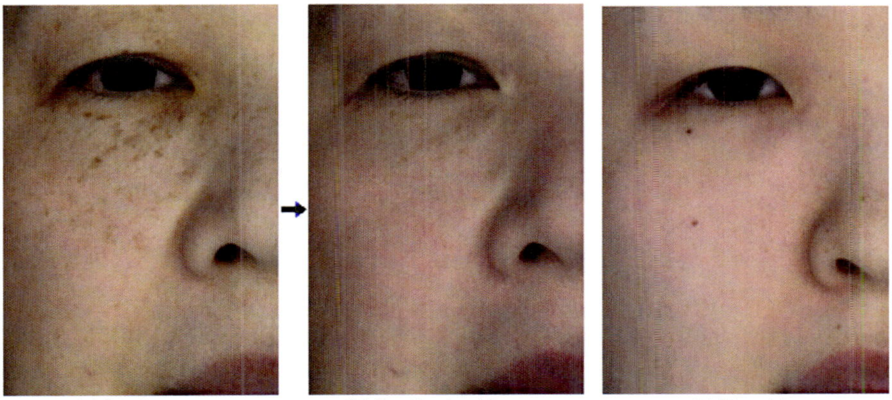

Fig. 3.21 Skin texture synthesis: (a) Original image of a 50-year old female, (b) synthesized image of the 50-year old female with a 20-year old appearance generated using the method shown in Figure 3.18, and (c) the original image of a 20-year old female. (Courtesy of N. Tsumura; reprinted Figure 12 with permission [168]; (c)2003 ACM, Inc.)

Several techniques for representing skin texture have been proposed in the past. Tsumura et al. have proposed a method for realistic synthesis of textures such as freckles with ICA and a pyramid-based texture synthesis technique [168]. Figures 3.21(a) and 3.21(c) show skin images of a 50-year old woman and her 20-year old daughter. The melanin images were first separated from the original images using ICA, and then histograms of the distribution of melanin were represented using a pyramid. By matching the histograms at each level of the pyramid, a synthetic image of the 50-year old woman with a 20-year old appearance was generated, as shown in Figure 3.21(b).

Haro et al. have developed a method for rendering 3D textured faces by directly measuring the geometrically fine structure of the skin surface [70]. In order to capture the fine structure, they took imprints from skin and then computed the surface normals using a shape from shading algorithm. By using the computed surface normals and measured BRDF, the synthetic face shown in the right column of Figure 3.22 was generated. By comparing this textured skin to skin without texture (the left column of Figure 3.22), we see that the fine details of the skin surface add realism to the rendered faces. For instance, specular

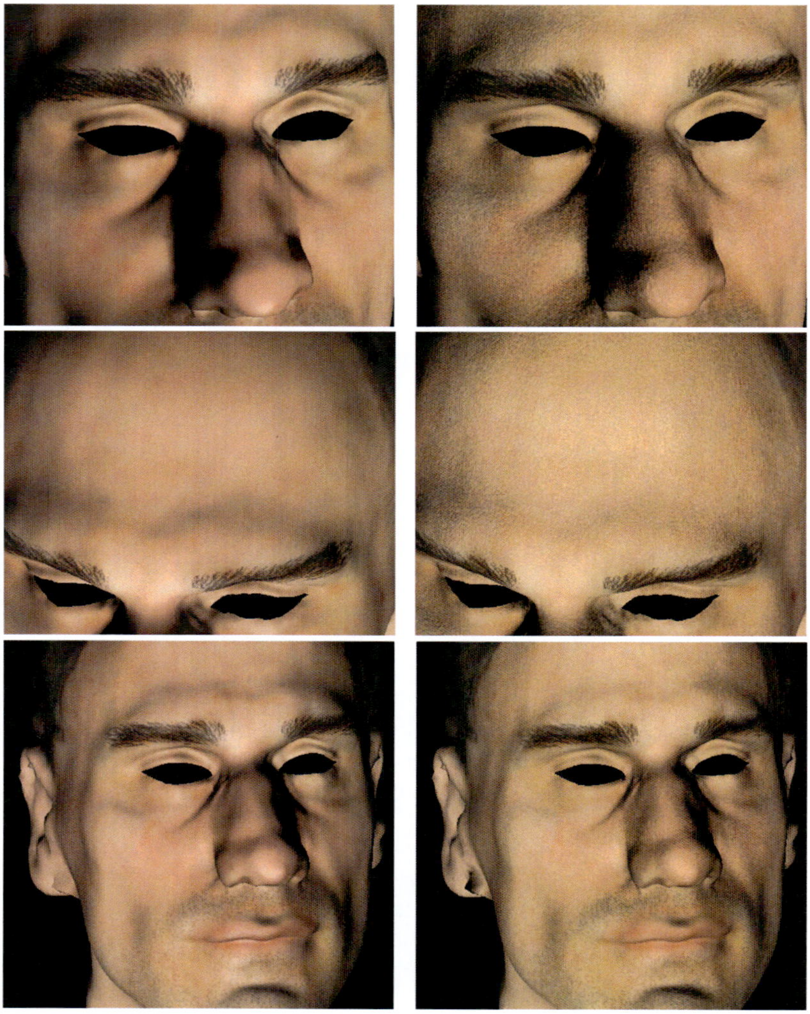

Fig. 3.22 Skin rendered without fine-scale structure (the left column) and with fine-scale structure (the right column). The appearance of skin with fine-scale structure appears more realistic. (Courtesy of A. Haro; reprinted Figure 7 with permission [70]; ©2001 Springer-Verlag.)

highlights such as those on the forehead and the left cheek look more realistic when there is fine detail within the highlights.

The appearance of skin texture varies significantly depending on the direction from which it is viewed and illuminated. Figure 3.23 shows

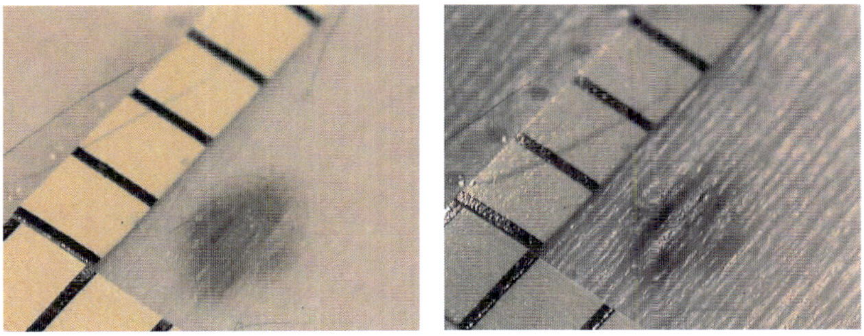

Fig. 3.23 Images of the same skin surface obtained with frontal (left) and oblique (right) illumination for the same viewing direction. The appearance of skin texture is strongly affected by the illumination direction. While color variations are apparent in the left image, the geometric details are clearer in the right image. (Courtesy of O. Cula and K. Dana.)

an example of the directional variation of skin texture. The appearance of the texture varies dramatically with the lighting direction. While frontal illumination brings out the color variations, more oblique lighting makes the surface details more visible. Figure 3.24 shows another example. Here, the change in viewing direction gives rise to the visible effects of occlusion, foreshortening, etc.

The above directional variations due to illumination and viewing are caused by geometrical phenomena such as shadowing and interreflection. Conventional two-dimensional approaches for representing skin texture [23, 70, 80, 127] cannot describe such directional variations.

Fig. 3.24 Images of the same skin surface obtained with a frontal camera view (left) and an oblique view (right) under the same illumination. The appearance of skin texture is also affected by the viewing condition. (Courtesy of O. Cula and K. Dana.)

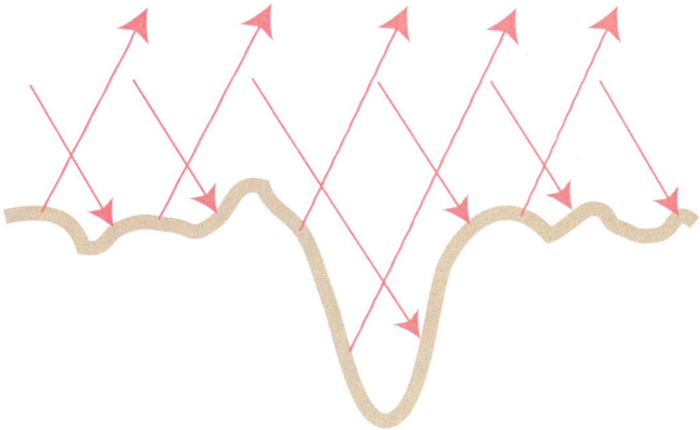

Fig. 3.25 Illustration of the BTF of skin. The BTF captures the surface details of skin, such as wrinkles and pores, as a function of viewing and illumination directions.

The Bidirectional Texture Function (BTF) [39] is a useful concept for modeling the directional variation of the appearance of texture. Figure 3.25 illustrates the optical phenomena underlying the BTF. Similar to the BRDF that describes the directional variation of reflectance at a surface point, the BTF models the directional variation of texture as a function of four imaging angles (viewing and illumination directions). A detailed study of the BTF of skin has been conducted by Cula et al. [34, 36, 37].

In order to build a database of skin BTF, Cula et al. captured images of skin surfaces with a CCD camera while changing the viewing and illumination directions. Skin texture tends to vary according to body region, age, and gender. Therefore, the images were obtained for different body regions (the cheek, nose, forehead, and chin on the face, a fingertip, a bottom segment on the palm, and the back side of the index finger) of both females and males whose ages range from 24 to 53 years. Figures 3.26 and 3.27 are examples (from facial regions and fingers) from the acquired BTF database. This database is unique as it includes high resolution images of various types of skin textures and can help further research on skin texture. The full database is available at http://www.caip.rutgers.edu/rutgers_texture [35].

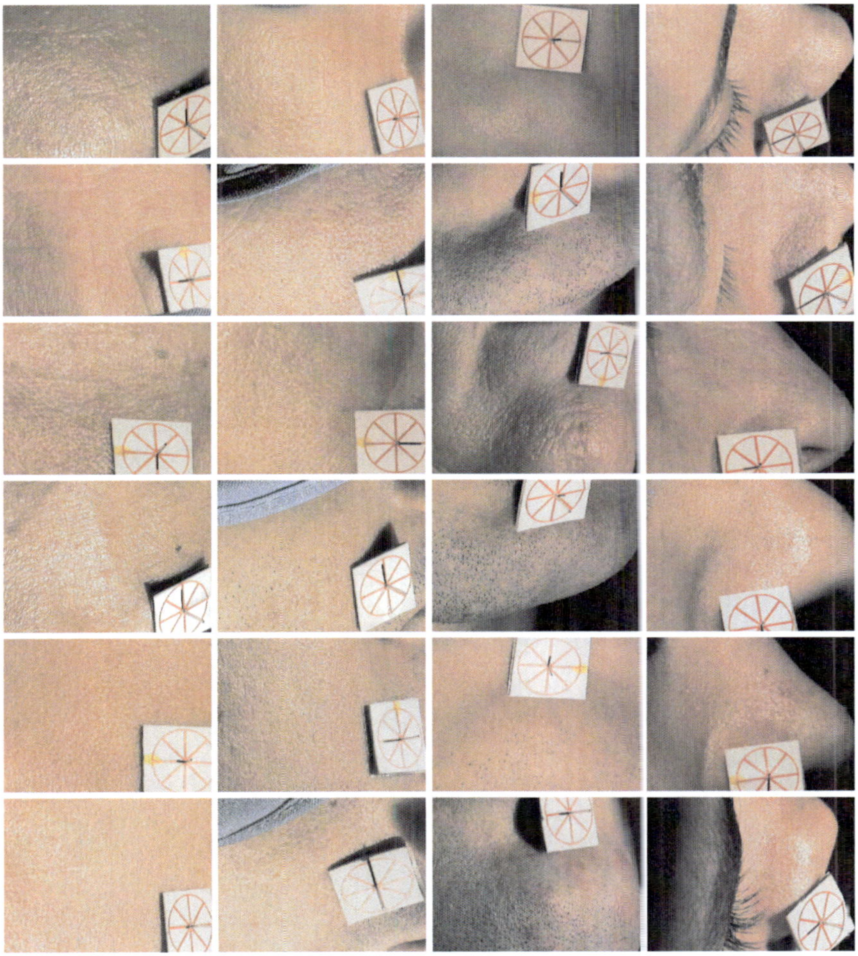

Fig. 3.26 Examples of facial skin texture images from http://www.caip.rutgers.edu/rutgers_texture [35]. Each row contains images from a certain human subject, while the columns correspond to the forehead, cheek, chin, and nose (left to right). These examples clearly show the differences between skin textures at different locations of the same subject and at same locations across different subjects. (Courtesy of O. Cula and K. Dana.)

3.5 Appearance of Body Regions and Body Parts

The appearance of skin varies across different body regions. This is because the optical and visual properties of skin we reviewed in the previous sections usually have regional variations. Such variations are

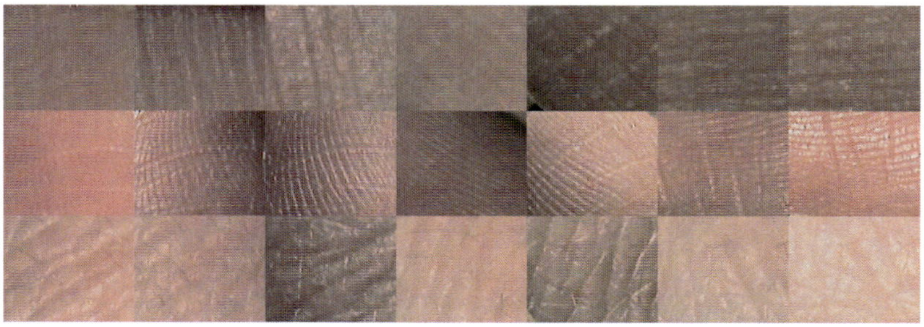

Fig. 3.27 Examples of skin texture images obtained from the index fingers of different subjects. First row shows images of the bottom segment on the palm side, the second row shows images of the fingertip, and the third row shows images of the bottom segment on the back side. (Courtesy of O. Cula and K. Dana.)

very important in some applications. For example, in computer graphics, when a face is rendered, the differences in the optical properties between the different facial parts (i.e., forehead, nose, cheek, chin, etc.) must be taken into account. Otherwise, the resulting image will look unnaturally homogeneous. In order to understand such regional variations, it is important to systematically tabulate the scattering and absorption coefficients, the BRDF/BSSRDF, and the BTF of various body regions and body parts. Unfortunately, there are very few studies in this direction.

The regional variations of scattering and absorption coefficients have been reported in several previous studies [46, 85, 91, 102, 106]. For example, the scattering and absorption coefficients in fingertips tend to be lower than those in the forearm [106]. The dermis of the leg has less scattering than abdominal dermis [85]. In Table 3.4, the scattering and absorption coefficients of the arm, foot sole, and forehead are given. This data clearly shows that the scattering and absorption coefficients vary across the body. These variations are primarily due to physio-anatomical differences in the components in Levels 1 and 2, such as the thickness and density of collagen fibers, the structure of the skin layers (e.g., thickness), the water content in the layers, etc.

The BRDF/BSSRDF of skin layers varies across the body because the regional variation of scattering and absorption properties influence light transport in the skin layers. For example, the spatial span of the distribution of reflected light due to subsurface scattering in skin layers becomes smaller in regions with high scattering coefficients. The appearance of such regions tends to be more opaque compared to a body region that has a lower scattering coefficient. For instance, Jensen et al. have reported that the arm skin tends to be more translucent than facial skin [91].

The BRDF/BSSRDF of skin is also influenced by the regional variations of SSLF, hairs, and fine wrinkles. These variations can be modeled by considering their physio-anatomical variations. In the case of facial skin, for example, the appearance of the T-zone tends to be shinier than that of other regions. This is mainly because the T-zone has thicker SSLF than other facial parts. The appearance of the cheek is sometimes softer and more matte than that of the nose. This is because the cheek usually has more vellus hairs and fine wrinkles, which induce asperity scattering and diffuse reflection, respectively.

The BTF of skin also varies depending on the body part, as seen in Figure 3.26. We also know that body regions that frequently change their morphologies (e.g., regions around eyes and mouth, forehead) tend to have deep wrinkles. There are usually more pores in the T-zone since there are more hair follicles compared to other facial regions. The skin texture database compiled by Cula and Dana [35] provides some insights into the variation of skin appearance with respect to body regions/parts.

4

Summary

One of the major goals of skin research is to develop a precise and yet computationally efficient model of skin appearance. In this survey, we have seen that many important results have been obtained that bring us closer to achieving this goal. However, we have also seen that many important and challenging problems remain unsolved and need to be addressed before we can reach this goal.

First, we still need to compactly describe the compound effect of the numerous optical phenomena that are simultaneously induced by the various skin components. Most of the previous studies focus on only one or two optical phenomena in isolation. For example, faces rendered using BRDF and BSSRDF models reviewed in Section 3.3 succeed in capturing the directional properties of reflection but cannot handle important skin features. Furthermore, some of the works only focus on the reflection properties of skin layers and ignore optical effects induced by the SSLF, hairs, and fine wrinkles, which are almost always present on top of the skin layers. In short, skin appearance is the end result of many different optical phenomena and much work needs to be done before we can have a unified and efficient computational model of skin appearance.

To describe skin appearance in any given setting, we first need to know which of the components of skin are at work. From this viewpoint, we hope our taxonomy in Figure 1.3 will prove useful. For example, to render a face (part appearance, Level 6), we should consider the optical properties of the components in the lower levels (i.e., Levels 1 to 5) such as the BRDF, BSSRDF, BTF, and region appearance. In order to describe the reflection properties of skin layers (BRDF/BSSRDF in Level 3), scattering and absorption properties of the skin layers (Levels 1 and 2) should be taken into account. This is because the reflection properties of the skin layers, especially subsurface scattering, depend on their components' scattering and absorption properties.

In addition, inter- and intra-individual variations of skin appearance should be studied in detail. This is especially important for realistic rendering of humans in computer graphics. For instance, as mentioned in Section 3.5, the variation of appearance over the face (nose, cheek, forehead, etc.) should be taken into account while rendering a face. Also, it is important to account for inter-person variation of skin appearance since no two persons have exactly the same skin over the entire body. For this reason, it would be very useful to collect data related to the optical and visual properties corresponding to Levels 1 to 4 of a large number of people.

Over the years, the study of skin appearance has become increasingly relevant to many fields, including, medicine, cosmetology, computer graphics, and computer vision. We believe this trend will continue for many years to come. We hope this survey will serve as a useful resource for future research on skin appearance.

Acknowledgments

We are grateful to Anne Fleming for patiently reading through early drafts of this survey and providing valuable comments.

We thank the following individuals and organizations for providing many of the figures and images included in this survey: C. Shea, Duke University Medical Center; T.L. Ray, University of Iowa College of Medicine; A.M. Cohen, Tel-Aviv University; G. Imokawa, Kao Corporation; S.L. Jacques, Oregon Health and Science University; S.R. Marschner, Cornell University; P. Hanrahan, Stanford University; J. Stam, Alias/Wavefront; S. Pont, Utrecht University; H.W. Jensen, University of California, San Diego; N. Tsumura, Chiba University; E. Claridge, University of Birmingham; A. Haro, Georgia Institute of Technology; O. Cula and K. Dana, Rutgers University; Nanoworld Image Gallery, Centre for Microscopy and Micro Nanoanalysis, The University of Queensland; Kao Corporation.

We also thank the following publishers for granting permission to use some of the figures in the report. Figure 2.3 is from B. Alberts, D. Bray, A. Johnson, J. Lewis, M. Raff, K. Roberts and P. Walter, *Essential Cell Biology* p601 ©1998 Garland Science Publishing Reprinted with permission; Figure 2.8 is from K.P. Wilhelm, P. Elsner,

E. Berardesca and H.I. Maibach, *Bioengineering of the Skin: Skin Surface Imaging and Analysis*, p202 and 205, ©1997 CRC Press Reprinted with permission; Figure 2.10 is from S. Akazaki, H. Nakagawa, H. Kazama, O. Osanai, M. Kawai, Y. Takema and G. Imokawa, *British Journal of Dermatology*, 147, 689, 2002 ©2002 Blackwell Publishing Reprinted with permission; Figure 3.10 is from P. Hanrahan and W. Krueger, *Proceedings of the 20th Annual Conference on Computer Graphics and Interactive Techniques SIGGRAPH 1993*, 165-174, 1993 ©1993 ACM, Inc. Reprinted with permission; Figure 3.11 is from J. Stam, *Proceedings of the 12th Eurographics Workshop on Rendering Techniques*, 39-52, 2001 ©2001 Springer-Verlag Reprinted with permission; Figure 3.13 and 3.14 are from J. Koenderink and S. Pont, *Machine Vision and Application; Special Issue on Human Modeling, Analysis and Synthesis*, 14, 260-268, 2003 ©2003 Springer-Verlag Reprinted with permission; Figure 3.16 is from H.W. Jensen, S.R. Marschner, M. Levoy and P. Hanrahan, *Proceedings of the 28th Annual Conference on Computer Graphics and Interactive Techniques SIGGRAPH 2001*, 511-518, 2001 ©2001 ACM, Inc. Reprinted with permission; Figure 3.17 is from N. Tsumura, H. Haneishi and Y. Miyake, *Journal of Optical Society of America A*, 16, 2169-2176, 1999 ©1999 Optical Society of America Reprinted with permission; Figure 3.18, 3.19 and 3.21 are from N. Tsumura, N. Ojima, K. Sato, M. Shiraishi, H. Shimizu, H. Nabeshima, S. Akazaki, K. Hori and Y. Miyake, *Proceedings of the 30th Annual Conference on Computer Graphics and Interactive Techniques SIGGRAPH 2003*, 770-779, 2003 ©2003 ACM, Inc. Reprinted with permission; Figure 3.20 is from E. Claridge, S. Cotton, P. Hall and M. Moncrieff, In *Medical Image Computing and Computer-assisted Intervention*, 730-738, 2002 ©2002 Springer-Verlag Reprinted with permission; Figure 38 is from A. Haro, B. Guenter and I. Essa, *Proceedings of the 12th Eurographics Workshop on Rendering Techniques*, 53-62, 2001 ©2001 Springer-Verlag Reprinted with permission.

References

[1] Adams Media Corporation Staff, *Milady's Standard: Textbook of Cosmetology*. Thomson Learning, 2000.

[2] S. Akazaki, H. Nakagawa, H. Kazama, O. Osanai, M. Kawai, Y. Takema, and G. Imokawa, "Age-related changes in skin wrinkles assessed by a novel three-dimensional morphometric analysis," *British Journal of Dermatology*, vol. 147, pp. 689–695, 2002.

[3] B. Alberts, D. Bray, A. Johnson, J. Lewis, M. Raff, K. Roberts, and P. Walter, *Essential Cell Biology*. pp. 601. Garland, 1998.

[4] P. H. Andersen, "Reflectance spectroscopic analysis of selected experimental dermatological models," *Skin Research and Technology*, vol. 3, pp. 8–15, 1997.

[5] R. R. Anderson, "Polarized light examination and photography of the skin," *Archives in Dermatology*, vol. 127, pp. 1000–1005, 1991.

[6] R. R. Anderson and J. A. Parrish, "The optics of human skin," *Journal of Investigative Dermatology*, vol. 77, pp. 13–19, 1981.

[7] R. R. Anderson and J. A. Parrish, *The Science of Photomedicine, Optical Properties of Human Skin*. ch. 6, Plenum Press, 1982.

[8] E. Angelopoulou, "The reflectance spectrum of human skin," Tech. Rep., University of Pennsylvania, 1999.

[9] E. Angelopoulou, "Understandking the color of human skin," in *Proceedings of the SPIE Conference on Human Vision and Electronic Imaging VI*, vol. 4299, pp. 243–251, 2001.

[10] E. Angelopoulou, "The uniqueness of the color of human skin," *Electronic imaging*, vol. 11, p. 5, 2001.

[11] E. Angelpopulou, R. Molana, and K. Daniilidis, "Multispectral skin color modeling," in *Proceedings of Computer Vision and Pattern Recognition: Technical Sketches*, pp. 635–642, 2001.

[12] E. Angelpopulou, R. Molana, and K. Daniilidis, "Multispectral skin color modeling," Tech. Rep., University of Pennsylvania, 2001.

[13] R. Aronson, "Boundary conditions for diffusion of light," *Journal of Optical Society of America A*, vol. 12, pp. 2532–2539, 1995.

[14] R. Aronson and N. Corngold, "Photon diffusion coefficient in an absorbing medium," *Journal of Optical Society of America A*, vol. 16, pp. 1066–1071, 1999.

[15] S. R. Arridge, M. Schweiger, M. Hiraoka, and D. T. Delpy, "A finite element approach for modeling photon transport in tissue," *Medical Physics*, vol. 20, pp. 299–309, 1993.

[16] V. Backman, R. Gurjar, K. Badizadegan, I. Itzkan, R. R. Dasari, L. T. Perelman, and M. S. Feld, "Polarized light scattering spectroscopy for quantitative measurement of epithelial cellular structures in situ," *IEEE Journal of Selected Topics in Quantum Electronics*, vol. 5, pp. 1019–1026, 1999.

[17] M. Bassani, F. Martelli, G. Zaccanti, and D. Contini, "Independence of the diffusion coefficient from absorption: Experimental and numerical evidence," *Optics Letters*, vol. 22, pp. 853–855, 1997.

[18] D. Batisse, R. Bazin, T. Baldeweck, B. Querleux, and J. L. Leveque, "Influence of age on the wrinkling capacities of skin," *Skin Research and Technology*, vol. 8, pp. 148–154, 2002.

[19] B. Beauvoit, T. Kitai, and B. Chance, "Contribution of the mitochondrial compartment to the optical properties of the rat liver: A theoretical and practical approach," *Biophysical Journal*, vol. 67, pp. 2501–2510, 1994.

[20] B. Beauvoit, T. Kitai, and B. Chance, "Time-resolved spectroscopy of mitochondria, cells and tissues under normal and pathological conditions," *Molecular and Cellular Biochemistry*, vol. 184, pp. 445–455, 1998.

[21] J. Beuthan, O. Minet, J. Helfmann, M. Herring, and G. Muller, "The spatial variation of the refractive index in biological cells," *Physics in Medicine and Biology*, vol. 41, pp. 369–382, 1996.

[22] F. Bevilacqua and C. Depeursinge, "Monte Carlo study of diffuse reflectance at sourcedetector separations close to one transport mean free path," *Journal of Optical Society of America*, vol. 16, pp. 2935–2945, 1999.

[23] L. Boissieux, G. Kiss, N. Thalman, and P. Kalra, "Simulation of skin aging and wrinkles with cosmetics insight," in *Europraphics Workshop on Animation and Simulation*, pp. 15–28, 2000.

[24] W. A. G. Bruls and J. C. van der Leun, "Forward scattering properties of human epidermal layers," *Photochemistry and Photobiology*, vol. 40, pp. 231–242, 1984.

[25] A. Brunsting and P. Mullaney, "Light scattering from coated spheres: Model for biological cells," *Applied Optics*, vol. 11, pp. 675–680, 1972.

[26] E. K. Chan, B. Sorg, D. Protsenko, M. O'Neil, M. Motamedi, and A. J. Welch, "Effects of compression on soft tissue optical properties," *IEEE Journal of Selected Topics in Quantum Electronics*, vol. 2, pp. 943–950, 1996.

[27] B. Chen, K. Stamnes, and J. J. Stamnes, "Validity of the diffusion approximation in bio-optical imaging," *Applied Optics*, vol. 40, pp. 6356–6366, 2001.

[28] J. H. Chung, "Photoaging in Asians," *Photodermatology Photoimmunology Phtomedicine*, vol. 19, pp. 109–121, 2003.

[29] E. Claridge, S. Cotton, P. Hall, and M. Moncrieff, "From colour to tissue histology: Physics based interpretation of images of pigmented skin lesions," in *Medical Image Computing and Computer-Assisted Intervention*, pp. 730–738, 2002.

[30] R. L. Cook and K. E. Torrance, "A reflectance model for computer graphics," *ACM Transactions on Graphics*, vol. 1, no. 1, pp. 7–24, 1982.

[31] P. Corcuff, J. L. Leveque, G. L. Grove, and A. M. Kligman, "The impact of aging on the microrelief of peri-orbital and leg skin," *Journal of The Society of Cosmetic Chemists*, vol. 82, pp. 145–152, 1987.

[32] S. D. Cotton, "Developing a predictive model of human skin colouring," in *Proceedings of SPIE Medical Imaging 1996*, vol. 2708, pp. 814–825, 1996.

[33] S. D. Cotton and E. Claridge, "Do all human skin colours lie on a defined surface within LMS space?," Tech. Rep. CSR-96-1, School of Computer Science, The Univeristy of Birmingham, 1996.

[34] O. G. Cula and K. J. Dana, "Image-based skin analysis," in *Proceedings of Texture 2002*, pp. 35–94, 2002.

[35] O. G. Cula, K. J. Dana, F. P. Murphy, and B. K. Rao, *Skin Texture Database at Rutgers University*. http://www.caip.rutgers.edu/rutgers_texture/cvg/index.html.

[36] O. G. Cula, K. J. Dana, F. P. Murphy, and B. K. Rao, "Bidirectional imaging and modeling of skin texture," *IEEE Transaction on Biomedical Engineering*, vol. 51, no. 12, pp. 2148–2159, 2004.

[37] O. G. Cula, K. J. Dana, F. P. Murphy, and B. K. Rao, "Skin texture modeling," *International Journal of Computer Vision*, vol. 62, no. 1–2, pp. 97–119, 2005.

[38] K. Dana, B. Ginneken, S. Nayar, and J. Koenderink, *Columbia-Utrecht Reflectance and Texture (CuReT) Database*. http://www1.cs.columbia.edu/CAVE/curet/.

[39] K. J. Dana, B. V. Ginneken, S. K. Nayar, and J. J. Koenderink, "Reflectance and texture of real-world surfaces," *ACM Transactions on Graphics*, vol. 18, no. 1, pp. 1–34, 1999.

[40] P. Debevec, T. Hawkins, C. Tchou, H.-P. Duiler, W. Sarokin, and M. Sagar, "Acquiring the reflectance field of a human face," in *Proceedings of ACM Transactions on Graphics, SIGGRAPH2000*, pp. 145–156, 2000.

[41] D. T. Delpy, M. Cope, P. van der Zee, S. Arridge, S. Wray, and J. Wyatt, "Estimation of optical pathlength through tissue from direct time of flight measurement," *Physics in Medicine and Biology*, vol. 33, pp. 1433–1442, 1988.

[42] S. G. Demos and R. R. Alfano, "Optical polarization imaging," *Applied Optics*, vol. 36, pp. 150–155, 1997.

[43] M. Denda and M. Takahashi, "Measurement of facial skin thickness by ultrasound method," *Journal of Society Cosmetic Chemists Japan*, vol. 23, pp. 316–319, 1990.

[44] C. Donner and H. W. Jensen, "Light diffusion in multi-layered translucent materials," *ACM Transactions on Graphics*, vol. 24, no. 3, pp. 1032–1039, 2005.
[45] C. Donner and H. W. Jensen, "A spectral BSSRDF for shading human skin," in *Eurographics Symposium on Rendering*, pp. 409–417, 2006.
[46] R. M. P. Doombos, R. Lang, M. C. Aalders, and H. J. C. M. Sterenborg, "The determination of in vivo human tissue optical properties and absolute chromophore concentrations using spatially resolved steady-state diffuse reflectance spectroscopy," *Physics in Medicine and Biology*, vol. 44, pp. 967–981, 1998.
[47] R. Drezek, A. Dunn, and R. R. Kortum, "Light scattering from cells: Finite-difference time-domain simulations and goniometric measurements," *Applied Optics*, vol. 38, pp. 3651–3661, 1999.
[48] R. Drezek, A. Dunn, and R. Richards-Kortum, "A pulsed finite-difference time-domain (FDTD) method for calculating light scattering from biological cells over broad wavelength ranges," *Optical Express*, pp. 147–157, 2000.
[49] A. Dunn, *Light Scattering Properties of Cells*. PhD thesis, University of Texas at Austin, 1997.
[50] A. Dunn and R. R. Kortum, "Three-dimensional computation of light scattering from cells," *IEEE Journal of Selected Topics in Quantum Electronics*, vol. 2, pp. 898–905, 1998.
[51] T. Durduran, D. A. Boas, B. Chance, and A. G. Yodh, "Validity of the difusion equation for small heterogeneities," *Advances in Optical Imaging and Photon Migration OSA Trends in Optics and Photonics Series*, vol. 2, pp. 60–63, 1996.
[52] T. Durduran, A. G. Yodh, B. Chance, and D. A. Boas, "Does the photon-diffusion coefficient depend on absorption?," *Journal of Optical Society of America*, vol. 14, pp. 3358–3365, 1997.
[53] D. J. Durian, "The diffusion coefficient depends on absorption," *Opticas Letters*, vol. 23, pp. 1502–1504, 1998.
[54] L. G. Farkas, ed., *Anthropometry of the Head and Face*. Raven Press, 1994.
[55] T. J. Farrell and M. S. Patterson, "A diffusion theory model of spatially resolved, steady-state diffuse reflectance for the noninvasive determination of tissue optical properties in vivo," *Medical Physics*, vol. 19, pp. 879—888, 1992.
[56] T. J. Farrell and M. S. Patterson, "Experimental verification of the effect of refractive index mismatch on the light fluence in a turbid medium," *Journal of Biomedical Optics*, vol. 6, pp. 468–473, 2001.
[57] T. J. Farrell, M. S. Patterson, and M. Essenpreis, "Influence of layered tissue architecture on estimates of tissue optical properties obtained from spatially resolved diffuse reflectometry," *Applied Optics*, vol. 37, pp. 1958–1972, 1998.
[58] J. F. Federici, N. Guzelsu, H. C. Lim, G. Jannuzzi, T. Findley, H. R. Chaudhry, and A. B. Ritter, "Noninvasive light-reflection technique for measuring soft-tissue stretch," *Applied Optics*, vol. 38, pp. 6653–6660, 1999.
[59] G. H. Findlay, "Blue skin," *British Journal of Dermatology*, vol. 83, pp. 127–134, 1970.
[60] K. Furutsu, "Diffusion equation derived from space-time transport equation," *Journal of Optical Society of America*, vol. 70, pp. 360–366, 1980.

[61] K. Furutsu, "Boundary conditions of the diffusion equation and applications," *Physical Review A*, vol. 39, pp. 1386–1401, 1989.

[62] K. Furutsu and Y. Yamada, "Diffusion approximation for a dissipative random medium and the applications," *Physical Review E*, vol. 50, pp. 3634–3640, 1994.

[63] M. Gniadecha and G. B. E. Jemec, "Quantitative evaluation of chronological ageing and photoageing in vivo: Studies on skin echogenicity and thickness," *British Journal of Dermatology*, vol. 139, pp. 815–821, 1998.

[64] L. Gobin, L. Blanchot, and H. Saint-Jalmes, "Integrating the digitized backscattered image to measure absorption and reduced-scattering coefficients in vivo," *Applied Optics*, vol. 38, pp. 4217–4227, 1999.

[65] R. Graaff, A. C. M. Dassel, M. H. Koelink, F. F. M. de Mul, J. G. Aarnoudse, and W. G. Zijlstra, "Optical properties in human dermis in vitro and in vivo," *Applied Optics*, vol. 32, pp. 435–447, 1993.

[66] R. Grover, O. Grobbelaar, B. D. G. Morgan, and D. T. Gault, "A quantitative method for the assessment of facial rejuvenation: A prospective study investigating the carbon dioxide laser," *British Journal of Plastic Surgery*, vol. 51, pp. 8–13, 1998.

[67] N. Guzlksu, J. F. Federici, H. C. Lim, H. R. Chauhdry, A. B. Ritter, and T. Findley, "Measurement of skin stretch via light reflection," *Journal of Biomedical Optics*, vol. 8, pp. 80–86, 2003.

[68] P. Hanrahan and W. Krueger, "Reflection from layered surfaces due to subsurface scattering," in *Proceedings of the 20th Annual Conference on Computer Graphics and Interactive Techniques SIGGRAPH1993*, pp. 165–174, 1993.

[69] J. D. Hardy, H. T. Hammell, and D. Murgatroyd, "Spectral transmittance and reflectance of excised human skin," *Journal of Applied Physiology*, vol. 9, pp. 257–264, 1956.

[70] A. Haro, B. Guenter, and I. Essa, "Real-time photo-realistic, physically based rendering of fine scale human skin structure," in *Proceedings of the 12th Eurographics Workshop on Rendering Techniques*, pp. 53–62, 2001.

[71] R. C. Haskell, L. O. Svaasand, T. T. Tsay, T. C. Feng, M. S. McAdams, and B. J. Tromberg, "Boundary conditions for the diffusion equation in radiative transfer," *Journal of Optical Society of America A*, vol. 11, pp. 2727–2741, 1994.

[72] J. Hata, M. Shimada, Y. Yamada, A. Uchida, M. Itoh, Y. Nakayama, and T. Yatagai, "Treatment of nevus using medical tattooing," *Journal of Biomedical Optics*, vol. 8, pp. 93–101, 2003.

[73] S. Hayashi, K. Mimura, and Y. Nishijima, "Changes in surface configuration of the skin caused by aging and application of cosmetics: Three-dimensional analysis according to a new system based on image analysis and fourier transformation," *International Journal of Cosmetic Science*, vol. 11, pp. 67–85, 1989.

[74] L. C. Henyey and J. L. Greenstein, "Diffuse radiation in the galaxy," *The Astrophysics Journal*, vol. 93, pp. 70–83, 1941.

[75] A. H. Hielscher, R. E. Alcouffe, and R. L. Barbour, "Comparison of finite-difference transport and diffusion calculations for photon migration in homo-

geneous and heterogeneous tissues," *Physics in Medicine and Biology*, vol. 43, pp. 1285–1302, 1998.

[76] A. H. Hielscher, S. L. Jacques, L. Wang, and F. K. Tittel, "The influence of boundary conditions on the accuracy of diffusion theory in time-resolved reflectance spectroscopy of biological tissues," *Physics in Medical Biology*, vol. 40, pp. 1957–1975, 1995.

[77] M. Hiraoka, M. Firband, M. Essenpreis, M. Cope, S. R. Arridge, P. van der Zee, and D. T. Delpy, "A monte carlo investigation of optical pathlength in inhomogeneous tissue and its application to near-infrared spectroscopy," *Physics in Medicine and Biology*, vol. 38, pp. 1859–1876, 1993.

[78] A. F. Hood, T. H. Kwan, M. C. Mihm Jr., T. D. Horn, and B. R. Smoller, *Primer of Dermatopathology*. Lippincott Williams & Wilkins, Third Edition, 2002.

[79] G. Imokawa, "Face washing and cleaning foams," *Fragrance Journal*, vol. 74, pp. 38–47, 1985.

[80] T. Ishii, T. Yasuda, S. Yokoi, and J. Toriwaki, "A generation model for human skin texture," *Communicating with Virtual Worlds*, pp. 139–150, 1993.

[81] S. L. Jacques, "Origins of tissue optical properties in the UVA, visible, and NIR regions," *Trends in Optics and Photonics: Advances in Optical Imaging and Photon Migration*, vol. 2, pp. 364–371, 1996.

[82] S. L. Jacques, "Optical absorption of melanin," *Oregon Medical Laser Center Monthly News and Articles on Biomedical Optics and Medical Lasers*, http://omlc.ogi.edu/spectra/ melanin/index.html, 1998.

[83] S. L. Jacques, "Video imaging with polarized light finds skin cancer margins not visible to dermatologists," *Oregon Medical Laser Center Monthly News and Articles on Biomedical Optics and Medical Lasers*, http://omlc.ogi.edu/news/feb98/polarization/index.html, 1998.

[84] S. L. Jacques, "Scattering of polarized light by biological tissues," *Oregon Medical Laser Center Monthly News and Articles on Biomedical Optics and Medical Lasers*, http://omlc.ogi.edu/news/oct99/saratov/index.htm, 1999.

[85] S. L. Jacques, C. A. Alter, and S. A. Prahl, "Angular dependence of HeNe laser light scattering by human dermis," *Lasers in the Life Science*, vol. 1, pp. 309–333, 1987.

[86] S. L. Jacques and S. A. Prahl, "Modeling optical and thermal distributions in tissue during laser irradiation," *Lasers in Surgery and Medicine*, vol. 6, pp. 494–503, 1987.

[87] S. L. Jacques, J. C. R. Roman, and K. Lee, "Imaging skin pathology with polarized light," *Journal of Biomedical Optics*, vol. 7, pp. 329–340, 2002.

[88] S. L. Jacques, J. R. Roman, and K. Lee, "Imaging superficial tissues with polarized light," *Lasers in Surgery and Medicine*, vol. 26, pp. 119–129, 2000.

[89] S. Jacques, "Skin optics," *Oregon Medical Laser Center Monthly News and Articles on Biomedical Optics and Medical Lasers*, http://omlc.ogi.edu/news/jan98/skinoptics.html, 1998.

[90] S. Jaspers, H. Hopermann, G. Saurermann, U. Hoppe, R. Lunderstadt, and J. Ennen, "Rapid in vivo measurement of the topography of human skin by

active image triangulation using a digital micromirror device," *Skin Research Technology*, vol. 5, pp. 195–207, 1999.

[91] H. W. Jensen, S. R. Marschner, M. Levoy, and P. Hanrahan, "A practical model for subsurface light transport," in *Proceedings of the 28th annual conference on Computer graphics and interactive techniques SIGGRAPH 2001*, pp. 511–518, 2001.

[92] M. Keijzer, S. L. Jacques, S. A. Prahl, and A. J. Welch, "Light distributions in artery tissue," *Lasers in Surgery and Medicine*, vol. 9, pp. 148–154, 1989.

[93] M. Keijzer, J. W. Pickering, and M. J. C. van Gemert, "Laser beam diameter for port wine stain treatment," *Lasers in Surgery and Medicine*, vol. 11, pp. 601–605, 1991.

[94] M. Keijzer, W. M. Star, and P. R. M. Storchi, "Optical diffusion in layered media," *Applied Optics*, vol. 27, pp. 1820–1824, 1988.

[95] A. Kienle, L. Lilge, M. S. Patterson, R. Hibst, R. Steiner, and B. C. Wilson, "Spatially resolved absolute diffuse reflectance measurements for noninvasive determination of the optical scattering and absorption coefficients of biological tissue," *Applied Optics*, vol. 35, pp. 2304–2314, 1996.

[96] A. Kienle and M. S. Patterson, "Improved solutions of the steady-state and the time-resolved diffusion equations for reflectance from a semi-infinite turbid medium," *Journal of Optical Society of America A*, vol. 14, 1997.

[97] A. Kienle, M. S. Patterson, N. Dgnitz, R. Bays, G. Wagnires, and H. van den Bergh, "Noninvasive determination of the optical properties of two-layered turbid media," *Applied Optics*, vol. 37, pp. 779–791, 1998.

[98] A. Kim and A. Ishimaru, "Optical diffusion of continuous-wave, pulsed, and density waves in scattering media and comparisons with radiative transfer," *Applied Optics*, vol. 37, pp. 5313–5319, 1998.

[99] M. Kobayashi, Y. Ito, N. Sakauchi, I. Oda, I. Konishi, and Y. Tsunazawa, "Analysis of nonlinear relation for skin hemoglobin imaging," *Optics Express*, vol. 9, pp. 802–812, 2001.

[100] J. Koenderink and S. Pont, "The secret of velvety skin," *Machine Vision and Applications; Special Issue on Human Modeling, Analysis and Synthesis*, vol. 14, pp. 260–268, 2003.

[101] J. S. Koh, H. Kang, S. W. Choi, and H. O. Kim, "Cigarette smoking associated with premature facial wrinkling: Image analysis of facial skin replicas," *International Journal of Dermatology*, vol. 41, pp. 21–27, 2002.

[102] H. Kolarova, D. Ditrichov, and J. Wagner, "Penetration of the laser light into the skin in vitro," *Lasers in Surgery and Medicine*, vol. 24, pp. 231–235, 1999.

[103] N. Kollias, "The physical basis of skin colour and its evaluation," *Clinical Dermatology*, vol. 13, pp. 361–367, 1995.

[104] A. Krishnaswamy and G. V. G. Baranoski, "A biophysically-based spectral model of light interaction with human skin," in *Eurographics*, vol. 23, pp. 331–340, 2004.

[105] G. Kumar and J. M. Schmitt, "Micro-optical properties of tissue," in *Proceedings of SPIE: Advances in Laser and Light Spectroscopy to Diagnosis Cancer and Other Diseases 3: Optical Biopsy*, vol. 2679, pp. 106–116, 1996.

[106] M. Larsson, H. Nilsson, and T. Stromberg, "In vivo determination of local skin optical properties and photon path length by use of spatially resolved diffuse reflectance with applications in laser doppler flowmetry," *Applied Optics*, vol. 42, pp. 124–134, 2003.

[107] C. Lasagni and S. Seidenari, "Echographic assessment of age-dependent variations of skin thickness," *Skin Research and Technology*, vol. 1, pp. 81–85, 1995.

[108] M. Lees, *Skin Care: Beyond the Basis*. Milady, 2001.

[109] J. L. Leveque, "EEMCO guidance for the assessment of skin topography," *Journal of European Academic Dermatol*, vol. 12, pp. 103–114, 1999.

[110] J. L. Leveque and P. Corcuff, *Noninvasive Methods for the Quantification of Skin Functions*. Springer, 1993.

[111] W. F. Lever and G. Schaumburg-Lever, *Histopathology of the Skin*. J. B. Lippincott Company, Seventh Edition, 1990.

[112] J. Q. Lu, X. H. Hu, and K. Dong, "Modeling of the rough-interface effect on a converging light beam propagating in a skin tissue phantom," *Applied Optics*, vol. 39, pp. 5890–5897, 2000.

[113] R. Marchesini, A. Bertoni, S. Andreola, E. Melloni, and A. E. Sichirollo, "Extinction and absorption coefficients and scattering phase functions of human tissues in vitro," *Applied Optics*, vol. 28, pp. 2318–2324, 1989.

[114] P. Marquet, F. Bevilacqua, C. Depeursinge, and E. B. de Haller, "Determination of reduced scattering and absorption coefficients by a single charge-coupled-device array measurement, part 1: Comparison between experiments and simulations," *Optical Engineering*, vol. 34, pp. 2055–2062, 1995.

[115] S. R. Marschner, S. H. Westin, E. P. F. Lafortune, K. E. Torrance, and D. P. Greenberg, "Image-based BRDF measurement including human skin," in *Proceedings of 10th Eurographics Workshop on Rendering*, pp. 139–152, 1999.

[116] S. R. Marschner, S. H. Westin, E. P. F. Lafortune, K. E. Torrance, and D. P. Greenberg, "Reflectance measurement of human skin," Tech. Rep., Cornell University, 1999.

[117] M. L. Mello, B. C. Vidal, A. C. de Carvalho, and A. C. Caseiro-Filho, "Change with age of anisotropic properties of collagen bundles," *Gerontology*, vol. 25, pp. 2–8, 1979.

[118] J. Mobley and T. Vo-Dinh, *Biomedical Photonics Handbook*. CRC Press, 2003.

[119] S. P. Morgan, M. P. Khong, and M. G. Somekh, "Effects of polarization state and scatterer concentration on optical imaging through scattering media," *Applied Optics*, vol. 36, pp. 1560–1565, 1997.

[120] S. P. Morgan and M. E. Ridgmay, "Polarization properties of light backscattered from a two layer scattering medium," *Optics Express*, vol. 7, pp. 395–402, 2000.

[121] M. Motamedi, S. Rastegar, G. LeCarpentier, and A. J. Welch, "Light and temperature distribution in laser irradiated tissue: The influence of anisotropic scattering and refractive index," *Applied Optics*, vol. 28, pp. 2230–2237, 1989.

[122] J. R. Mourant, M. Canpolat, C. Brocker, O. Esponda-Ramos, T. M. Johnson, A. Matanock, K. Stetter, and J. P. Freyer, "Light scattering from cells: The

contribution of the nucleus and the effects of proliferative status," *Journal of Biomedical Optics*, vol. 5, pp. 131–137, 2000.
[123] J. R. Mourant, J. P. Freyer, A. H. Hielscher, A. A. Eick, D. Shen, and T. M. Johnson, "Mechanisims of light scattering from biological cells relecant to noninvasive optical-tissue diagnosis," *Applied Optics*, vol. 37, pp. 3586–3593, 1998.
[124] J. R. Mourant, T. M. Johnson, and J. P. Freyer, "Characterizing mammalian cells and cell phantoms by polarized backscattering fiber-optic measurements," *Applied Optics*, vol. 40, pp. 5114–5123, 2001.
[125] H. Murakami, T. Horii, N. Tsumura, and Y. Miyake, "Measurement and simulation of 3D gonio spectral reflectance of skin surface," *Digital Biocolor Journal*, pp. 2.1–2.6, 2002.
[126] R. Murphy and D. W. K. Cotton, "Computer-assisted image analysis of skin surface replicas," *British Journal of Dermatology*, vol. 124, pp. 571–575, 1991.
[127] M. Nahas, H. Huitric, M. Rioux, and J. Domey, "Facial image synthesis using skin texture recording," *The Visual Computer*, vol. 6, pp. 337–343, 1990.
[128] H. Nakai, Y. Manabe, and S. Inokuchi, "Simulation and analysis of spectral distributions of human skin," in *Proceedings of 14th International Conference of Pattern Recognition*, vol. 2, pp. 1065–1067, 1998.
[129] S. K. Nayar and M. Oren, "Generalization of the lambertian model and implications for machine vision," *International Journal of Computer Vision*, vol. 14, pp. 227–251, 1995.
[130] E. J. Naylor, "The structure of the cornea as revealed by polarized light," *Quarterly Journal of Microsopical Science*, vol. 94, pp. 83–88, 1953.
[131] S. Nickell, M. Hermann, M. Essenpreis, T. J. Farrell, U. Kramer, and M. S. Patterson, "Anisotropy of light propagation in human skin," *Physics in Medicine and Biology*, vol. 45, pp. 2873–2886, 2000.
[132] F. E. Nicodemus, J. C. Richmond, J. J. Hsia, I. W. Ginsberg, and T. Limperis, *Geometric Considerations and Nomenclature for Reflectance*. National Bureau of Standards (US), 1977.
[133] A. Nilsson, P. Alsholm, A. Karlsson, and S. Andresson-Engels, "T-matrix computations of light scattering by red blood cells," *Applied Optics*, vol. 37, pp. 2735–2748, 1998.
[134] H. Nilsson, M. Larsson, G. E. Nilsson, and T. Stromberg, "Photon pathlength determination based on spatially resolved diffuse reflectance," *Journal of Biomedical Optics*, vol. 7, pp. 478–485, 2002.
[135] Y. Nomura, O. Hazeki, and M. Tamura, "Exponential attenuation of light along nonlinear path," *Advanced Experimental Medical Biology*, vol. 248, pp. 77–80, 1989.
[136] Y. Nomura, O. Hazeki, and M. Tamura, "Relationship between time-resolved and non-time-resolved Beer-Lambert law in turbid media," *Physics in Medicine and Biology*, vol. 42, pp. 1009–1022, 1997.
[137] L. O. Olsen, H. Takiwaki, and J. Serup, "Skin thickness and echographic density of 22 anatomical sites," *Skin Research and Technology*, vol. 1, pp. 74–82, 1995.

[138] M. Ooe and H. Shiroshita, "The relationship between transparent skin and optical properties of stratum corneum," *Fragrance Journal*, vol. 4, pp. 38–44, 2002.

[139] L. T. Perelman, J. Wu, I. Itzkan, and M. S. Feld, "Photon migration in turbid media using path integrals," *Physical Review letters*, vol. 72, pp. 1341–1344, 1994.

[140] C. Pierard-Franchimont and G. E. Pierard, "Assessment of aging and actinic damages by cyanoarcylate skin surface strippings," *American Journal of Dermapathology*, vol. 9, pp. 500–507, 1987.

[141] S. Prahl, "Optical absorption of hemoglobin," *Oregon Medical Laser Center Monthly News and Articles on Biomedical Optics and Medical Lasers*, http://omlc.ogi.edu/spectra/hemoglobin/index.html, 1998.

[142] S. A. Prahl, *Light Transport in Tissue.* PhD thesis, University of Texas at Austin, 1988.

[143] S. A. Prahl, M. J. C. van Gemert, and A. J. Welch, "Determining the optical properties of turbid media by using the adding-doubling method," *Applied Optics*, vol. 32, pp. 559–568, 1993.

[144] P. T. Pugliese, *Physiology of the Skin II.* ch.1, p. 1, Allured Publishing Corporation, 2001.

[145] M. H. Ross and L. J. Romrell, *Histology a Text and Atlas.* Williams and Wilkins, 1989.

[146] I. S. Saidi, S. L. Jacques, and F. K. Tittel, "Mie and Rayleigh modeling of visible-light scattering in neonatal skin," *Applied Optics*, vol. 34, pp. 7410–7418, 1995.

[147] F. A. Schellander and J. T. Headington, "The stratum corneum: some structural and functional correlates," *British Journal of Dermatology*, vol. 91, pp. 507–515, 1974.

[148] R. J. Scheuplein, "A survey of some fundamental aspects of the absorption and reflection of light by tissue," *Journal of Society of Cosmetic Chemists*, vol. 15, pp. 111–122, 1964.

[149] J. Schmitt and G. Kumar, "Optical scattering properties of soft tissue: A discrete particle model," *Applied Optics*, vol. 37, pp. 2788–2797, 1998.

[150] J. M. Schmitt, G. X. Zhou, and E. C. Walker, "Multilayer model of photon diffusion in skin," *Journal of Optical Society of America A*, vol. 7, pp. 2141–2153, 1990.

[151] H. M. Sheu, S. C. Chao, T. W. Wong, Y. Y. Lee, and J. C. Tsai, "Human skin surface lipid film: An ultrastructural study and interaction with corneocytes and intercellular lipid lamellae of the stratum corneum," *British Journal of Dermatology*, vol. 140, pp. 385–391, 1999.

[152] M. Shimada, Y. Masuda, Y. Yamada, M. Itoh, M. Takahashi, and T. Yatagai, "Explanation of human skin color by multiple linear regression analysis based on the modified Lambert-Beer law," *Optical Review*, vol. 7, pp. 348–352, 2000.

[153] C. R. Simpson, M. Kohl, M. Essenpreis, and M. Cope, "Near-infrared optical properties of *ex vivo* human skin and subcutaneous tissues measured using the Monte Carlo inversion technique," *Physics in Medicine and Biology*, vol. 43, pp. 2465–2478, 1998.

[154] C. So-Ling and L. Ling, "A multi-layered reflection model of natural human skin," in *Proceedings of Computer Graphics International 2001*, pp. 249–256, 2001.

[155] K. Sokolov, R. Drezek, K. Gossage, and R. Richards-Kortum, "Reflectance spectroscopy with polarized light: Is it sensitive to cellular and nuclear morphology," *Optical Express*, vol. 5, pp. 302–317, 1999.

[156] J. Stam, "An illumination model for a skin layer bounded by rough surfaces," in *Proceedings of the 12th Eurographics Workshop on Rendering*, pp. 39–52, 2001.

[157] I. M. Stockford, S. P. Morgan, P. C. Y. Chang, and J. G. Walker, "Analysis of the spatial distribution of polarized light backscattered from layered scattering media," *Journal of Biomedical Optics*, vol. 7, pp. 313–320, 2002.

[158] G. Streekstra, A. Hoekstra, E. Nijhof, and R. Heethaar, "Light scattering by red blood cells in ektacytometry: Fraunhofer versus anomalous diffraction," *Applied Optics*, vol. 32, pp. 2266–2272, 1993.

[159] G. J. Streekstra, A. G. Hoekstra, E. J. Nijhof, and R. M. Heethaar, "Anomalous diffraction by arbitarily oriented ellipsoids: Applications in ektacytometry," *Applied Optics*, vol. 33, pp. 7288–7296, 1994.

[160] S. Takatani and M. D. Graham, "Theoretical analysis of diffuse reflectance from a two-layer tissue model," *IEEE Transactions on Biomedical Engineering*, vol. 26, pp. 656–664, 1979.

[161] H. Takiwaki, Y. Kanno, Y. Miyaoka, and S. Arase, "Computer simulation of skin color based on a multilayered skin model," *Skin Research and Technology*, vol. 3, pp. 36–41, 1997.

[162] K. Torrance and E. Sparrow, "Theory for off-specular reflection from roughened surfaces," *Journal of Optical Society of America*, no. 57, pp. 1105–1114, 1967.

[163] D. Toublanc, "Henyey-Greenstein and Mie phase functions in Monte Carlo radiative transfer computations," *Applied Optics*, vol. 5, no. 18, pp. 3270–3274, 1996.

[164] C. Tsai, Y. F. Yang, C. C. Han, J. H. Hsieh, and M. Chang, "Measurement and simulation of light distribution in biological tissues," *Applied Optics*, vol. 40, pp. 5770–5777, 2001.

[165] K. Tsukahara, Y. Takema, T. Fujimura, S. Moriwaki, and M. Hattori, "Quantitative two-dimensional analysis of facial wrinkles of Japanese women at various ages," *International Journal of Cosmetic Science*, vol. 24, pp. 71–80, 2002.

[166] N. Tsumura, H. Haneish, and Y. Miyake, "Independent component analysis of spectral absorbance image in human skin," *Optical Review*, vol. 7, pp. 479–482, 2000.

[167] N. Tsumura, H. Haneishi, and Y. Miyake, "Independent component analysis of skin color image," *Journal of Optical Society of America A*, vol. 16, pp. 2169–2176, 1999.

[168] N. Tsumura, N. Ojima, K. Sato, M. Shiraishi, H. Shimizu, H. Nabeshima, S. Akazaki, K. Hori, and Y. Miyake, "Image-based skin color and texture analysis/synthesis by extracting hemoglobin and melanin information in the

skin," in *Proceedings of the 30th Annual Conference on Computer Graphics and Interactive Technicques SIGGRAPH 2003*, pp. 770–779, 2003.

[169] N. Tsumura, K. Uetsuki, N. Ojima, and Y. Miyake, "Correlation map analysis between appearance of japanese facial images and amount of melanin and hemoglobin components in the skin," in *Proceedings of SPIE: Human Vision and Electronic Imaging 6*, vol. 4299, pp. 252–260, 2001.

[170] V. V. Tuchin, "Light scattering study of tissue," *Physics-Uspekhi*, vol. 40, pp. 495–515, 1997.

[171] V. V. Tuchin, S. R. Utz, and I. V. Yaroslavsky, "Tissue optics, light distribution, and spectroscopy," *Optical Engineering*, vol. 33, pp. 3178–3188, 1994.

[172] T. Uchida, T. Komeda, M. Miyagi, H. Koyama, and H. Funakubo, "Quantification of skin aging by three-dimensional measurement of skin surface contour," in *Systems, Man, and Cybernetics, 1996, IEEE International Conference*, vol. 1, pp. 450–455, 1996.

[173] M. J. C. van Gemert, S. L. Jacques, H. J. C. M. Sterenborg, and W. M. Star, "Skin optics," *IEEE Transaction on Biomedical Engineering*, vol. 36, pp. 1146–1154, 1989.

[174] W. Verkruysse, G. W. Lucassen, and M. J. C. van Gemert, "Simulation of color of port wine stain: Skin and its dependence on skin variables," *Lasers in Surgery and Medicine*, vol. 25, pp. 131–139, 1999.

[175] G. Videen and D. Ngo, "Light scattering multipole solution for a cell," *Journal of Biomedical Optics*, vol. 3, pp. 212–220, 1998.

[176] S. Wan, R. R. Anderson, and J. A. Parish, "Analytical modeling for the optical properties of the skin with in vitro and in vivo appplications," *Photochemistry and Photobiology*, vol. 34, pp. 493–499, 1981.

[177] L. Wang and S. L. Jacques, *Monte Carlo Modeling of Light Transport in Multi-layered Tissues in Standard C*. PhD thesis, University of Texas, M. D. Anderson Cancer Center, 1992.

[178] L. Wang, S. L. Jacques, and L. Zheng, "MCML — Monte Carlo modeling of light transport in multi-layered tissues," *Computer Methods and Programs in Biomedicine*, vol. 47, pp. 131–146, 1995.

[179] L. Wang, S. L. Jacques, and L. Zheng, "CONV–convolution for responses to a finite diameter photon beam indident on multi-layered tissues," *Computer Methods and Programs in Biomedicine*, vol. 54, pp. 141–150, 1997.

[180] L. H. Wang and S. L. Jacques, "Source of error in calculation of optical diffuse reflectance from turbid media using diffusion theory," *Computer Methods and Programs in Biomedicine*, vol. 61, pp. 163–170, 2000.

[181] A. J. Welch, G. Yoon, and M. J. C. van Gemert, "Practical models for light distribution in laser-irradiated tissue," *Lasers in Surgery and Medicine*, vol. 6, pp. 488–493, 1987.

[182] A. J. Weltch and M. J. C. van Gemert, eds., *Optical Thermal Response of Laser-Irradiated Tissue*. Plenum Press, 1995.

[183] T. Weyrich, W. Matusik, H. Pfister, B. Bickel, C. Donner, C. Tu, J. McAndless, J. Lee, A. Ngan, H. W. Jensen, and M. Gross, "Analysis of human faces using a measurement-based skin reflectance model," *ACM Transactions on Graphics*, vol. 25, pp. 1013–1024, 2006.

[184] K. Wilhelm, P. Elsner, E. Berardesca, and H. I. Maibach, eds., *Bioengineering of the Skin: Skin Surface Imaging and Analysis*. CRC Press, 1997.

[185] B. C. Wilson and G. Adam, "A Monte Carlo model for the absorption and flux distributions of light in tissue," *Medical Physics*, vol. 10, pp. 824–834, 1983.

[186] Y. Wu, P. Kalra, and N. M. Thalmann, "Simulation of static and dynamic wrinkles of skin," *The Visual Computer*, vol. 15, pp. 183–198, 1999.

[187] M. A. Wverett, E. Yeargers, R. M. Sayre, and R. L. Olson, "Penetration of epidermis by ultraviolet rays," *Photochemistry and Photobiology*, vol. 5, pp. 533–542, 1966.

[188] Y. E. Yarker, R. M. Aspden, and D. W. L. Hukins, "Birefringence of articular cartilage and the distribution of collagen fibril orientations," *Connective Tissue Research*, vol. 11, pp. 207–213, 1983.

[189] A. N. Yaroslavskaya, S. R. Utz, S. N. Tatarinstev, and V. V. Tuchin, "Angular scattering properties of human epidermal layers," in *Cell and Biotissue Optics: Application in Laser Diagnostics and Therapy*, vol. 2100, pp. 38–41, 1994.

[190] K. M. Yoo, F. Liu, and R. R. Alfano, "When does the diffusion approximation fail to describe photon transport in random media?," *Physical Review Letters*, vol. 64, pp. 2647–2650, 1990.

[191] G. Yoon, S. A. Prahl, and A. J. Welch, "Accuracies of the diffusion approximation and its similarity relations for laser irradiated biological media," *Applied Optics*, vol. 28, pp. 2250–2255, 1989.

[192] G. Yoon, A. L. Welch, M. Motamedi, and M. C. J. van Gemert, "Development and application of three-dimensional light distribution model for laser irradiated tissue," *IEEE Journal of Quantum Electronics*, vol. QE-23, pp. 1721–1733, 1987.

[193] A. R. Young, "Chromophores in human skin," *Physics in Medicine and Biology*, vol. 42, pp. 789–802, 1997.

[194] W. G. Zijlstra, A. Buursma, and W. P. Meeuwsen-van der Roset, "Absorption spectra of human fetal and adult oxyhemoglobin, de-oxyhemoglobin, carboxyhemoglobin, and methemchlobin," *Clinical Chemistry*, vol. 37, pp. 1633–1638, 1991.

[195] G. Zonios, J. Bykowski, and N. Kollias, "Skin melanin, hemoglobin, and light scattering properties can be quantitatively assessed in vivo using diffuse reflectance spectroscopy," *Journal of Investigative Dermatology*, vol. 117, pp. 1452–1457, 2001.